WORSE THAN GLOBAL WARMING: WAVE TECHNOLOGY

The imminent threat to civilization as told by the prophecies.

NINA ANDERSON

WORSE THAN GLOBAL WARMING:
WAVE TECHNOLOGY

by
NINA ANDERSON

Copyright© 2007 by Nina Anderson

All Rights Reserved.

No part of this book may be reproduced in any form without the written consent of the publisher.

Printed in the USA

ISBN 978-1-884820-86-1
LCC # 2007906089

1. Environment 2. New Age 3. Metaphysics 4. Science 5. Astronomy

Safe Goods/ ATN Publishing
561 Shunpike Rd. • Sheffield, MA 01257
413-229-7935
www.safegoodspub.com

cover photograph:
Roman Krochuk, Fairbanks, AK, "Northern Lights"

printed on recycled paper

Preface

Readers of my recent novel, *2012 Airborne Prophesy*, have asked that I elaborate on the technology described within that story line and questioned the end days prophecies surrounding the year 2012. They wanted to know if it was fact or fiction, and voiced concerns that the deadly implications levied by the main characters in the book could actually befall our present day civilization and create another apocalypse that is not related directly to global warming.

Unfortunately, I must report that all reference to the myriad of current electromagnetic weapons described in *2012 Airborne Prophesy* is based on actual research. Not only do the weapons described exist but they are and have been in use for some time now. Our biggest known threat to the survival of planet Earth has been the nuclear bomb. We are now exceeding that singular threat tenfold. Not only are we looking at extinction from man-induced global warming and environmental abuse but we are falling victim to the unseen evil – planetary electromagnetic stress. Non-lethal weapons are currently being "tried" in Iraq. These weapons fire microwaves at humans, burning the skin as a deterrent. They could easily be converted to emit electromagnetic frequencies that are tuned to brain waves for the purpose of implanting subliminal messaging. Electromagnetic pulses are being used to control the weather and, in doing so, are triggering atmospheric imbalance. Under the guise of technology developed to improve communication, pulsed electromagnetic frequencies bombard the ionosphere, creating fissures that allow deadly radiation into our atmosphere.

Cell towers are popping up like candy spewing their controversial health risks. Cell phones are the cigarettes of the new millennium. Quite the fad and the convenience, these electronic devices have not been proven absolutely safe. European studies more and more frequently are revealing significant health risks. Could the cellular industry be a link to global brainwashing through the unseen frequency matrix? Perhaps this is considered science fiction, but who can prove that it won't happen. Rogue frequency waves caused by so many electronic devices and weapons could easily trigger a pole shift as the Atlanteans may have done 12,000 years ago that spawned hundreds of prophecies warning us that the end days are coming soon.

Worse Than Global Warming: Wave Technology was written specifically to draw the obvious parallels between the purported Atlantean scoiety and our current civilization. The similarities concerning use of technology, social pathways and political aims are unnerving. According to archeologists and researchers, the Ancients did possess a high level of civilization that became their downfall, possibly causing cataclysmic changes on Earth. If the findings prove valid, shouldn't we heed history? Following in their footsteps will only recreate the same outcome. It's as if we are caught in a time loop. Twelve thousand years later we are repeating the same mistakes. As the year 2012 draws near and the Mayan Calendar ends will we see a new, enlightened era as many prophets predict … or will a new doomsday win out because we triggered the "end days"?

Table of Contents

Introduction	7
Chapter I: *Reality Check*	11
Chapter II: *The Ancients*	15
Chapter III: *The Correlation*	27
Chapter IV: *Energy Waves Leading Us To Destruction*	45
Chapter V: *Accelerating Mother Nature's Rage*	59
Chapter VI: *Wave Technology: The big threat to modern civilization*	69
Chapter VII: *Can We Change The Prophecies?*	79
Epilogue	91
Bibliography	92

Introduction:
Global Warming

Until recently the news media has been focused on wars and civil disruptions worldwide. According to scientific calculations, about 15,000 wars have taken place on the planet during the recent five thousand years. More than 3.5 billion people have been killed in those wars. There were only 292 peaceful years in the entire history of the mankind. There is no reason to believe that something will change in the future. And to that note, mankind is developing highly sophisticated weaponry to assure they are victorious. As you will discover during the following pages, the new "star wars" weapons are mostly electronic in design and fill the air surrounding our planet with new frequencies. The effect this type of electronic wave saturation will have on the stability of the globe is unknown, and little is being done to study the long-term side effects prior to the introduction of these weapons into action. This is by far a greater threat to the survival of civilization as we know it than is global warming.

Former U.S. Vice President Al Gore's campaign to expose our influence on the environment has been met with quite a media frenzy. Whole magazine issues have focused on the subject, while threatened industry expends great amounts of dollars to challenge the reports and convince the public that global warming is not happening. There may be validity to their claims that CO^2 emissions are not the largest contributor to the planet's temperature increase. Actually, Earth does go through cycles of warming and cooling on a regular basis. To what degree humans are enhancing that temperature shift gives fuel to the debate.

We do know that the Earth is changing. "Five of the 10 hottest U.S. years on record occurred before 1940; three were in the past decade."[1] Desertification is causing 60,000 square kilometers of

---

[1] Jacoby, Jeff, "The jury is still out on global warming," The Boston Globe, Aug. 20, 2007

productive land to be lost each year.² Over a recent 10-year period, the average winter temperature on the Arctic ice cap shot up by seven degrees. Permafrost is melting causing houses to crumble off their foundations and creating problems for pipelines. Chunks of ice continue to slide into the Antarctic waters. According to Gore, "Almost all of the mountain glaciers in the world are now melting, many of them quite rapidly. There is a message in this."³ Tornadoes are on the increase. Hurricanes are getting stronger due to warmer ocean waters. Floods from severe monsoon seasons are plaguing Asia. As far as the bad rap U.S. CO^2 emissions have been getting we have a close competitor. If China's growth continues unchecked, the country will spew out five times as much CO^2 in the next twenty-five years as the Kyoto Protocol will save. Because of the focused attention on fear of destroying our planet, new technology is spawning alternatives to many of the man-induced perpetrators.

Worldwide, the auto industry is becoming obsessed with hybrids. Ethanol from corn is being viewed as the star alternative fuel. Unfortunately, it is purported to cost more and tax the environment more during its production. Plus, since farm production is gearing up for fueling autos, the feedlots will suffer – no food for the cows. Biodiesel from leftover frying oil, sewage, used tires and plastic bottles can all be transformed into fuel. Wind power is gearing up with tall turbines now gracing the landscape. Even units without blades are starting to appear with homebuilt rooftop kits available for the average citizen. Solar is experiencing a renewed interest. Artificial floating wetland archipelagos are being formed to restore coastal areas lost to housing. Cooling pumps have been designed by Phil Kithil from New Mexico to float beneath warm ocean waters to cool them, in effect reducing the size of surface waves, with the object being to reduce hurricane intensities. Peter Flynn of the University of Alberta Canada proposes a way of increasing

---

² Pickert, Kate, "The World Turns To Desert," *Popular Science*, Aug. 2007
³ Gore, Al, *An Inconvenient Truth*, Rodale, NY, NY. 2006

Arctic ice through giant ice-cube makers that use sprayed seawater. These would help stabilize the subsiding salt water diluted from melting glaciers and maintain the temperate climate of Europe generated from the Gulf Stream. Even the computer industry is getting on the bandwagon by considering replacing petroleum-filled cases with plant-based polymers, using solar power for laptops and instigating a recycling program for parts.[4]

But is mankind to blame or is global warming a natural phenomenon? Al Gore's, *An Inconvenient Truth,* states that "Today's climate pattern has existed throughout the entire history of human civilization." How far back do the researchers go? What about the cultures that existed before the last ice age thought to be 10,000 years ago? Recent evidence has shown that humans with technology existed long before that. Actually, ice ages occur every 250 *million* years … or, wait. every 250 *thousand* years … or maybe every 100 *thousand* years … or perhaps the real number is 10 *thousand* years.

Scientists still can't agree on the vital statistics of ice ages. Archeological evidence shows mammals in places where we thought there were only oceans and vice-versa. It seems that planet Earth has done flip-flops of the temperate zones over and over again. Ice ages seem to vary from north to south. The Arctic Circle shows abundant fossil evidence for subtropical conditions that once existed. There are glacial markings in Brazil. What would cause the ice to move?

One theory is magnetic pole reversal. Our continents have made passage through the polar places. One experiencing an ice age in Africa may be far removed from their relatives enjoying the beach in Finland during the same time period. The first leader of the IGY South Pole Station, Dr. Paul Siple, sought to explain the shifting of the poles by a "certain oscillation of the earth itself."[5] The earth's regions traverse the magnetic poles in due course according to the law of oscillation.

---

[4] *Popular Science* special issue: "The Future of the Environment," August, 2007
[5] Martinez, Susan, PhD, "Ice Age Fact or Fiction?" *Atlantis Rising*, No. 65

Reversed polarity is a natural mechanism. We are well aware of magnetic north being on the move. Airports in the last several years have had to have their runways renamed as their magnetic headings change.

So whom do we believe? Is global warming a natural trend that has happened before? Is it a function of planetary oscillation? Are we accelerating the natural order of things through our disregard for the environment? Or, will we intercept Mother's Nature's plan by throwing off her pulse?

If all the prophecies were warning that man cannot change and will repeat history, a strong message appears. Consider that the Earth is a living being: pulsing, breathing, belching, restless and harmonious. The American Indians cherished the land and thanked the plants and animals before chowing down. They knew if they mistreated the Earth it would not sustain them. Their prophecies reflected their fears. Not only are we cutting the veins and polluting the lungs of the Earth, but we are ravaging her nervous system. Humans with a malfunctioning heartbeat may face a catastrophic end to their physical being. How can we not see that by assaulting the pulse of the planet, she may react – and not in favor of our lifestyles. What is tickling her? Is the momentum for extreme weather a factor of global warming or energy waves caused by frequency saturation? The ancient prophecies may be right. We are entering a cosmic asteroid field (the dark rift) that exists in a portion of the Milky Way. In December, 2012, our solar system will be in a specific alignment that may exert a strong pull on Earth's stability. Our electromagnetic pulses may just be the straw that broke the camel's back and cause a weak and wobbling planet to flip as it may have done 12,000 years ago. Let's look at the speculated Ancient civilization and see similarities to our own present disregard for this sustainable spot in the heavens, consider their ignorance and heed their warnings.

Chapter I:
Reality Check

Enter into this story aboard a jet aircraft (tail number 9242X) as its two pilots, Ron and Kate, encounter an invisible anomaly:

"9242 X-ray requesting route deviation." I called to the controller without even consulting my Captain, Ron.
"9242 X-ray, what's happening up there?" the radio squawked.
"Radar's painting severe weather ahead. Like to deviate to the north, 42X."
"9242 X-ray, we show no weather on our radar screen and we are unable to grant deviation authority," the controllers responded. They were obviously unhappy with our request.
Ron pushed the talk button and said, "I don't care what the regulations say, we're not flying into that thing."
Kate and Ron looked ahead and saw absolutely no clouds to confirm what the radar showed as a very evil storm. The controller's radar depicted only clear skies, but the radar in the airplane didn't lie.
By now the radar echo was only ten miles in front of us and our radio reception was starting to break up. We could barely hear the controller. The autopilot was having difficulty tracking the Flight Management System input and when we selected it to track our new course, it flew us right through to a heading of 010 degrees. That was very unusual and at this moment we didn't need problems with our electronics.
We approached our descent altitude of 18,000 feet and still couldn't see any clouds. "The center of the weather on the radar is off to our left. I guess we dodged the bullet," I said to Ron, prematurely. Just then a

humming sound appeared to encase the cockpit, and the yoke started to vibrate.

"I'm disconnecting the autopilot, Kate. It may take two of us to keep this ship together so follow me through on the controls."

Red lights began to illuminate on the instrument panel. I urgently relayed to Ron the fact that we were losing electrical power, "My navigation instruments just went black. The electrical bus failure warning panel is lighting up like a Christmas tree and the generator fail lights are flickering." At that moment Ron also lost his instruments. I checked the circuit breakers and strangely none were popped. The engines were operating normally and the generators were still online, but we weren't getting any power to the electrical busses. Although we were having multiple electrical failures, there was no turbulence. We frantically ran through the emergency checklist but couldn't quite get a handle on what was the cause of the problems.

As we gained distance on the "cloud" the instruments started to come online. Ron and I were both relieved but wondered what had happened. We turned to our newly assigned heading and as the radar came up to speed again we tried to find the "storm" we had circumnavigated. It was gone!

Ron turned the airplane more to the south to see if we could pick up the echo, but nothing. "I'll be damned. Did we just enter the twilight zone or are both of us losing our minds?" he asked not really expecting a response. Clear air turbulence can give you quite a tumble without you seeing it coming but having a weather phenomena affect the aircraft systems was a new one to us.

This story was excerpted from *2012 Airborne Prophesy* and is a "reenactment" of a real life experience by a seasoned flight crew flying for a major corporation in the 1980s. The flight path took them over the Bermuda Triangle. Neither pilot believed in the stories surrounding disappearances over that part of the ocean, until that day. Upon landing, they both headed for the nearest pub to try to forget this unexplained event.

Could this anomaly have resulted from the Great Crystal that powered the Atlantean civilization and sunk beneath the ocean along with that vast continent? It is suggested that such an anomaly could be created when suboceanic plates shift causing the energy of the crystal to become activated. These energized pulses have been thought to cause dematerialization, severe weather and disrupt electronics due to their frequency output. If my pilot friends had flown into this radar-depicted invisible storm, would they have become part of the Bermuda Triangle legends of aircraft vanishing in thin air?

As we will see in subsequent chapters, modern scientists have created similar technology that could also cause this kind of anomaly. Maybe it was merely a test of a new defense weapon that materialized offshore as a radar return, as was discovered by the characters in *2012 Airborne Prophesy*.

Chapter II:
The Ancients

Many researchers allude to the fact that our present day civilization was not the first to have technology. Images found on the ceiling beams of a 3,000-year old New Kingdom Temple located several hundred miles south of Cairo and the Giza Plateau of Abydos reveal startling similarities to present day (and future) aircraft. A petroglyph from the Hathor Temple at Dendera near Abydos depicts what appears to be Egyptian figures holding electrical devices similar to a light bulb on a column with a cord connected to a box. In his book, *Investigating the Unexplained*, British scientist Ivan T. Sanderson explains that the column seems to be an insulator or the electrical generating device itself.[6]

Another depiction from an 18th dynasty papyrus scroll is interpreted as an orb on a device taken to be a static generator similar to the modern Van De Graaff generator found in many high school science laboratories. In such a device, static electricity builds up in the orb causing it to light up. At the Bogota Gold Museum, an artifact resembling a small delta-wing jet aircraft is on display. These are thought to be at least 1,000 years old. According to some research archivists the Ancients also had water pumps, cranes, hoists, coin operated machines, computers, radio and television.

Robots are not new. According to historian Andrew Tomas in *We Are Not the First*[7], the engineers of Alexandria in Greece had over 100 different automatons well over 2,000 years ago. Plato says that his robots were so active that they had to be prevented from running away. The ancient Chinese had bronze dragons that wagged their tails and the monk Gerbert d' Aurillac (920-1003), professor at the University of Rheims, was reported to have possessed a bronze robot that answered questions through a computer advising him on matters of politics and religion.

---
[1] Sanderson, Ivan T., *Investigating the Unexplained*, 1972, Prentice Hall, Englewood Cliffs, NJ
[2] Tomas, Andrew, *We Are Not The First*, 1971, Souvenir Press, London, UK

The Ancient civilizations with advanced technology spanned the globe. Reference has been made to an advanced technology in archived texts from China, India, the Middle East, South America and even the desert southwest of the USA. A zoomorphic pendant from Panama resembles a backhoe. In 1938, while conducting a dig at Khujut Rabu´a, Iraq, archeologist Dr. Wilhelm Koenig discovered an object that looked like today's dry-cell battery. Other similar finds surfaced. A woodcut of the legendary Chi-Kung people depicts them in flying machines. In the recently published ancient Sanskrit text, Samarangana Sutradhara of Bhoja, India, a whole chapter is dedicated to the principles of construction of various aircraft and other machines. In 1875, the Vaimanika Sastra, a fourth-century BC text written by Bharadvajy the Wise, using even older texts as his source, was rediscovered in a temple in India. It dealt with the operation of the Indian aircraft (Vimanas) and included information on steering, lightning and storm protection and how to switch the drive to "solar energy" from a free energy source similar to antigravity. This document has been translated into English.

According to ancient Ethiopian tradition as recorded in the Kebra Negast[8] a passage refers to air travel. "King Solomon would visit Makeda and his son Menelik by flying in a heavenly car. The king – and all who obeyed his word flew on the wagon without pain and suffering, and without sweat or exhaustion, and traveled in one day a distance which took three months to traverse (on foot)."

Proof of the theory of an atomic war comes from an examination of many locations where rock was vitrified. In northern India excavations at Mohenjodaro and Harappa revealed skeletons that were thousands of years old and had dropped in their tracks. One site measured the radio-activity of the skeletons and found them fifty times greater than normal, on par with those from Nagasaki and Hiroshima. Furthermore, thousands of clay vessels were shattered into fused "black stones" melted by extreme heat.

---

[8] Budge, Sire E.A. Wallis (translator), *The Queen of Sheba and Her Only Son Menyelek* (Kebra Nagast), 1932, Dover, London

In another city between the Ganges and the mountains of Rajmahal, huge masses of the walls of ancient structures are fused together, literally turned to glass! Since there were no indicated volcanic eruptions the only conjecture is an atomic blast or other type of heat weapon.

Even California's Death Valley gives indication that it was subject to an atomic attack. In 1850, Captain Ives William Walker viewed these ruins.[9] At the center there was a thirty-foot rock and the remains of a structure on top of it. The southern side of both the rock and the building was melted and vitrified. Tectonic heat was not a plausible explanation, nor volcanic activity so that left speculation as to atomic warfare. Scientists have since tried to find this city, as have many prospectors looking for gold from the "lost city."

Numerous ancient texts have taken the mystery out of history. If you correctly interpret them they tell of a time of peace and tranquility followed by oppression, war and ultimately an apocalypse. Two principle forces seemed to oppose each other in the years just prior to the earth changes that destroyed Atlantis. The Rama society (India) was comprised of a peace loving people who equated to a new-age utopian society. They had achieved a high level of technology, but used it for peaceful purpose until they had to defend themselves. On the other hand, the Atlanteans developed into power mongers and set out to push their values on the world. They, too, had a highly technological society, but had sunk into the realm of violence, power and greed. The Ramayana, Mahabarata and other texts speak of the hideous war that took place between these two factions, some twelve thousand years ago. Weapons of mass destruction (WMDs) were unleashed. Former readers of these texts could not comprehend such firepower until recent times revealed similar present day WMDs.

The ancient Mahabharata[10] relates the destructiveness of the war. "– (the weapon was) a single projectile charged with all the

---

[9] Noorbergen, Renee, *Secrets of the Lost Races*, 1977, Barnes & Noble Publishers, NY

power of the Universe. An incandescent column of smoke and flame as bright as the thousand suns rose in all its splendor – an iron thunderbolt, a gigantic messenger of death, which reduced to ashes the entire race of the Vrishnis and the Andhakas – the corpses were so burned as to be unrecognizable. The hair and nails fell out. Pottery broke without apparent cause, and the birds turned white – after a few hours all foodstuffs were infected – to escape from this fire the soldiers threw themselves in streams to wash themselves and their equipment."

It would appear what they described was an atomic bomb. The Russian archeologist, A. Gorbovsky, in his book *Riddles of Ancient History*,[11] mentions a high incidence of radiation in the skeletons found at the Indian sites of Mohenjodaro and Harappa. References like this one are not isolated. Indian texts describe many more battles that saw the use of a fantastic array of weapons and aerial vehicles. An ancient earthquake machine similar to the one Tesla invented was described as Scalar Wave Weaponry and had the potential to destroy far away civilizations by levying an electromagnetic pulse through the earth. It is uncanny that shortly after unleashing this massive weapon and the "nuclear bombs" the Ancients experienced a pole shift and apocalypse eliminating the continent of Atlantis.

Power Systems:

The use of crystals as part of a power generating system was a form of technology inherited from the culture of Atlantis lost after the polar shift and magnetic reversal 12,000 years ago. In the older Atlantean civilization, crystals were used quite extensively.[12] It is suspected that they were part of an elaborate system within pyramids that generated an ionized beam to the ionosphere creating an electrical charge that manifested a lightning bolt. This energy would be transmitted by standing electromagnetic waves

---

[10] Roy, Chandra Protap (translator), *The Mahabharata*, 1889, Calcutta, India
[11] A. Gorbovsky, *Riddles of Ancient History*, 1066, Soviet Publishers, Moscow

created from the lightning to step-down receivers that in turn sent energy to personal receivers to light houses and run equipment and to power vehicles. A global grid system was thought to be in operation at that time whereby there were generating stations in Atlantis, Egypt, Peru, England, Easter Island, India, Africa, west Africa, Japan, Australia, Russia, Brazil and Canada. These supplied power to the world. Drawing on this plan, modern day scientist Nikola Tesla proposed a similar wireless power grid with nine main transmitter locations using his invention, the Tesla coil, to generate a focused electric field. This device created an ionized beam that was used as a conductor to the ionosphere in effect creating a lightning strike that could be converted into power in the same vein as the Atlantean technology. His tests in Colorado Springs confirmed the viability of such a device and he claimed to have wirelessly transmitted 10,000 watts over a distance of 26 miles within the near field of the transmitter. Through this device he also created a standing electromagnetic wave that encircled the globe.

Numerous psychics and seers have described the Atlantean technology of crystal power. The modern Hindu sage and seer, Lama-Sing stated, "The Atlanteans used the knowledge of the crystal refraction, amplification and storage. It is known that a beam of light directed intensely and focused specifically on certain series of facets in a gem will, when it exits from the reflective plane of the gem, be amplified rather than diminished. And further, these amplified energies were broken down into a wide and sophisticated spectrum. The Atlantens used the spectrum of this energy so as to be more usable, and for a specific purpose, much as one would use petroleum in terms of its various spectrum limitations for specific purposes.

"Extracting from the same basic substance, they used certain divisions of the energy for growing things, for healing, and for knowledge. Other phases of spectrum for disassembling

---

[12] Jochmans, Joseph Robert, *Time Capsule: The Search for the Lost Hall of Records in Ancient Egypt,* Alma Tara Publishing, SC, 29731

molecular structures, and yet other combinations of these strata for building, assembling structures, as in chains; or producing matter, transmutation of matter and that sort. Their basic technology is still available in the earth plane in various locations. When it is proper and when it is in accord with God's will, you shall have it again. But with it will come a burden of spiritual decision and needed growth. Man cannot have hate, hostility, and anxiety within his being. He must replace this with love, tranquillity, compassion and patience. For at the time this knowledge is recovered, man will have many of the so-called secrets towards the creation and, conversely, the destruction of matter.

"Crystals have the ability to transfer energy, to retain it, to maintain its intensity, to focus and transmit it over great distance to similar receivers as are equal or comparable to the transmitter. Thus, from one pyramid to another the Atlanteans, in a sense, transmitted energy. That when the face of the earth, as it is called, was directed toward a certain point one pyramid would function to intensify and transmit energies to other pyramids which would then act as receiving devices and would disperse energy as it was needed. The opposite would be true, when that pyramid was at an unfocusable point to their celestial alignment the others would transmit to those. Very simple method, very effective method, though it brought them many difficulties later."[13]

Dr. Frank Alper, founder of the Arizona Metaphysical Society, in 1981 channeled series of readings dealing with life in ancient Atlantis. In reference to crystals he expounded, "The Atlantean crystals were natural forms, but their growths were speeded up. Some specimens of clear quartz were produced to almost twenty-five feet high and ten feet in diameter, had twelve sides, and were used for storing and transmitting power. Each had totally flawless and polished surfaces, so that there would be an undisturbed flow of energies through them from one crystal pole to the other. Smaller crystals were used for healing,

---

[13] Excerpted from "Crystal Power and the Energies in Atlantis," www.astraldelta.com

meditation, psychic development, increasing mental capacity, communications, power generators, dematerialization, and transport of objects, magnetic force fields, and travel at speeds only dreamed of by our culture today.

All these various crystals received their power from a variety of sources, including the Sun, the Earth's energy grid system, or from each other. The larger stones, called Fire Crystals, were the central receiving and broadcasting stations, while others acted as receivers for individual cities, buildings, vehicles and homes."[14]

One of the most detailed descriptions of the Atlantean use of a mysterious instrument called the Great Crystal, was given by the famed psychic of Virginia Beach, Edgar Cayce. "The Crystal was housed in a special building, oval in shape, with a dome that could be rolled back, exposing the Crystal to the light of the sun, moon and stars at the most favorable times. Atop the Crystal was a moveable capstone used to focus incoming rays of energy and direct current to various parts of the Atlantean countryside."

It appears the Crystal gathered solar, lunar, stellar, atmospheric and Earth energies as well as unknown elemental forces and concentrated these at a specific point, located between the top of the Crystal and the bottom of the capstone. Cayce mentioned that the Crystal became hot when in operation; it employed inducted methods; utilized a kind of wave energy other than electromagnetic; and it melted an invisible beam of energy that could pass through water and solid matter.

Another psychic who has received information on the mysterious power source of the Atlanteans is Washington columnist and author Ruth Montgomery. Montgomery learned that the ancient Atlanteans used an advanced energy created by what her spirit guides called the Great Crystal. The Atlanteans also built a reflector for it, and housed it in a gigantic domed building with a moveable top, so that the energy could be directed wherever desired. Nearby copper vats were constructed to store the energy. According to Montgomery, "The secret of the Great Crystal was

---

[14] ibid.

in its carbon structure, which by an unrevealed process was sufficiently powerful enough to raise the level of energy ten thousand times that of any known instrument today. It was also used as an important source of communication. The Atlanteans utilized certain Crystal frequencies to project sound and images, similar to our radio and television. The pictures were viewed on round images that projected a ray, and were thrown on walls or any blank surface, so that there was no need for screens or tubes.

"The Atlanteans also built temples that used the light spectrum from the sun's rays for the specific purpose of healing. The rays were directed to cubicles where patients stretched out on couches and received the beneficial properties without the dangerous ones."

Correspondingly, a considerable amount of research is being done today in the use of both low power and high power light waves for controlling cancer and skin ailments. In the Soviet Union, light beams with an output of only one three-thousandths that of a household 60-watt bulb have been discovered to stimulate skin rejuvenation and aid in skin grafting. In Hungary, Germany and Scandinavia, doctors are also successfully experimenting with "laser-puncture" or healing by the projection of light waves on body acupuncture points. And in Denmark, the Finsen Institute of Copenhagen has had remarkable results giving cancer patients light baths of green, blue and ultraviolet spectrum radiation.

Aircraft and weaponry:

Not long ago the Chinese discovered some Sanskrit documents in Lhasa, Tibet, and sent them to the University of Chandrigarh to be translated. Dr. Ruth Reyna of the University said recently that the documents contain directions for building interstellar spaceships! Their method of propulsion, she said, was "anti-gravitational" and was based upon a system analogous to that of "laghima," the unknown power of the ego existing in man's physiological makeup, "a centrifugal force strong enough to counteract all gravitational pull." According to Hindu Yogis, it is this "laghima" which enables a person to levitate. Dr. Reyna said that on board these machines (which were called Astras by the text) the ancient Indians could have sent a detachment of men onto any planet, according to the document, thought to be thousands of years old. The manuscripts were also said to reveal the secret of "antima, the cap of invisibility" and "garima, how to become as heavy as a mountain of lead." Naturally, Indian scientists did not take the texts very seriously until the Chinese announced that they were including certain parts of the data for study in their space program.

The manuscripts did not say definitely that interplanetary travel was ever made. However, one of the great Indian epics, the Ramayana, does have a highly detailed story in it of a trip to the moon in a Vimana (or Astra), and in fact details a battle on the moon with an Asvin (or Atlantean) airship.

The Rama Empire of Northern India and Pakistan developed at least fifteen thousand years ago on the Indian subcontinent and was a nation of many large, sophisticated cities, many of which are still to be found in the deserts of Pakistan, northern, and western India. Rama culture existed, apparently, parallel to the Atlantean and was ruled by enlightened Priest-Kings who governed the cities. According to ancient Indian texts, the people had flying machines that were called Vimanas. The ancient Indian epic describes a Vimana as a doubledeck, circular aircraft with portholes and a dome, much as we would imagine a flying saucer. It flew with the "speed of the wind" and gave forth a "melodious sound." There were at least four different types of Vimanas, some saucer shaped, others like long cylinders (cigar shaped airships). The ancient Indians, who manufactured these ships themselves, wrote entire flight manuals on the control of the various types of Vimanas, many of which are still in existence, with some even translated into English.

The Vaimanika Sastra (or Vymaanika-Shaastra) has eight chapters with diagrams, describing three types of aircraft, including details of the metal and propulsion systems. It also mentions thirty-one essential parts of these vehicles and sixteen types of materials from which they are constructed, which absorb light and heat. They were powered by some sort of anti-gravity, could take off vertically, and were capable of hovering in the sky, like a modern helicopter or dirigible. It describes pilot training and flying uniforms and a checklist of thirty-two instructions for the pilots to use prior to each flight.

Vimanas were kept in a Vimana Griha, a kind of hanger, and were sometimes said to be propelled by a yellowish-white liquid (gasoline?), and sometimes by some sort of mercury compound. According to the Dronaparva, part of the Mahabarata, and the Ramayana, one Vimana described was shaped like a sphere and born along at great speed on a mighty wind generated by mercury. It moved like a UFO, maneuvering up, down, backwards and forward, as the pilot desired. In another Indian source, the Samar, Vimanas were "iron machines, well-knit and smooth, with a charge of mercury that shot out of the back in the form of a roaring flame." It is theorized that mercury did contribute to the propulsion, or more possibly, the guidance system. Curiously, Soviet scientists have discovered what they call "age old instruments used in navigating cosmic vehicles" in caves in Turkestan and the Gobi Desert. The "devices" are described as hemispherical objects of glass or porcelain, ending in a cone with a drop of mercury inside.

It is evident that ancient Indians were global aviators. Writing found at Mohenjodaro in Pakistan and still undeciphered has also been found in one other place in the world: Easter Island. In the Mahavira of Bhavabhuti, a Jain text of the eighth century culled from older texts and traditions, we read, "an aerial chariot, the Pushpaka, conveys many people to the capital of Ayodhya. The sky is full of stupendous flying machines, dark as night, but picked out by lights with a yellowish glare."

According to the Indian texts, Atlanteans used their flying machines, "Vailixi," a similar type of aircraft, to literally try and subjugate the world. The Atlanteans, known as "Asvins" in the Indian writings, were apparently even more advanced technologically than the Indians, and certainly of a more war-like temperament. Although no ancient texts on Atlantean Vailixi are known to exist, some information has come down through esoteric, "occult" sources that describe their flying machines. Similar, if not identical to Vimanas, Vailixi were generally "cigar shaped" and had the capability of traveling underwater as well as in the atmosphere or even outer space. Other vehicles, like Vimanas, were saucer shaped, and could apparently also be submerged.

Eklal Kueshana, author of *The Ultimate Frontier*, wrote in 1966, that the Vailixi were first developed in Atlantis 20,000 years ago, and the most common ones are "saucer shaped of generally trapezoidal cross-section with three hemispherical engine pods on the underside. They use a mechanical antigravity device driven by engines developing approximately 80,000 horse power." Vailixi were also powered from the energy currents that originated from the Fire Crystals and traveled like a wave in the atmosphere.

We have not come close (at least publicly) to developing antigravity flying machines. Reports of flying saucers may not be idle imaginings as our governments may have accessed these ancient texts and followed their instructions. Secret antigravity or mercury vortex propulsion systems very well may currently be in operation. UFOs may not be alien spacecraft but simply a reconstruction of technology from the past. For details on potential construction of these futuristic craft from the past you may refer to the book *Atlantis and the Power Systems of the Gods*.[15] Authors, Childress and Clendenon do a thorough job of giving us technical data and reinforcing our speculative imaginations.

Weapons of Mass Destruction were in good supply 12,000 years ago. Not only did the Ancients possess nuclear firepower, but they also had weapons that could brainwash, deliver pathogens and vaporize targets. In the Mahabharatra, an ancient

---

[10] Childress, David Hatcher, & Clendenon, Bill, *Atlantis and the Power Systems of the Gods*, 2000, Adventures Unlimited, Kempton, IL

epic poem, lasers were referred to as "Indra's Dart" which operated via a circular reflector. When switched on it produced a "shaft of light" which when focused on a target, "consumed it with its power." They also had sound-seeking missiles as referred in the poem as "laid on an arrow which killed by seeking out sound." Could our military have access to the ancient WMD designs? There is an uncanny similarity in technological warfare between today and that of 12,000 years ago.

Chapter III:
The Correlation

The Power Grid:

Then: The Ancients are purported to have a distribution system for their power that utilized a unique and highly effective wireless technology. The Atlanteans used the knowledge of the crystal refraction, amplification and storage. When a beam of light directed intensely and focused specifically on certain series of the facets of the crystal, it will be amplified rather than diminished when it exits from the reflective plane. These energies could further be broken down into a wide and sophisticated spectrum for many uses. Stored crystal energy can be transmitted over great distances. In their system, generating stations used this source to transmit to sub receiver/local transmission generating stations. These energy stations were the recipients of power for distribution on a local level to be used for appliances, heating and lighting.

    The core generating station was theorized to be the Great Pyramid. This was a geomechanical power plant that initiated a focused electric field generated from an controlled hydrogen gas explosion. This field created an ionized beam aimed at the ionosphere for the purpose of generating lightning. The electrical current produced by this bolt would be received by a large copper capstone on top of the pyramid and stored in a crystal to be transmitted to receiver stations. The electromagnetic waves created from the lightning pulse circled the earth in two directions. As discovered by today's scientists, when this wave meets itself on the other side of the planet it reaches a crest in its amplitude creating a strong electric field between the surface of the earth and the ionosphere. If a receiver was placed at this location, power could have been generated for that area.

Another theory hypothesizes that crystals extracted energy from the earth responded sympathetically to the earth's vibrations and converted that energy into electricity. Crystals contract and expand when compressed and released becoming a transducer. Harmonics generated by the earth and amplified by resonators and the construction of the King's Chamber in the Great Pyramid had the ability to convert vibratory oscillations into electricity. Finely tuned crystals directed the energy from the Great Pyramid to satellite pyramids or obelisks that acted as receiving stations. Because they were free energy it did not seem possible to capitalize on any income stream from these power stations, a factor that keeps modern man from developing this type of technology.

The Unarius Academy of San Diego, California, theorizes about the Atlantean power system. The core of the main generating station was a huge rotating squirrel-cage generator that was housed in a twenty-foot square metal box on the floor just above the generator. A computer made and broke connections with banks of power collector cells on the outside pyramid surface in such a fashion that they created a great oscillating voltage. On top of the metal box was a ten-foot sphere that discharged electricity to a similar metal ball hanging down from the pyramid apex on a long metal rod. This discharge served as a tank-circuit creating sparks that jumped to another electrode creating necessary high-intensity voltage. On top of the pyramid was a "flagpole" with spokes that could send this voltage in a predetermined direction by employing finely tuned crystals. The net total of these charge and discharge oscillations was millions of megacycles per second at the rate of more than 186,000 miles per second, similar to our present day laser.[16] Maybe this device was not so much a power-generating station but a weapon.

Now: In the 20th century, Nikola Tesla (inventor of AC power) proposed a World System that included nine global transmitters at

---

[16] Childress, David Hatcher, & Clendenon, Bill, *Atlantis and the Power Systems of the Gods*, 2000, Adventures Unlimited, Kempton, IL

specific locations converting naturally-occurring, earth-based electromagnetism into power. He said we needed only ground contact to generate power. The electrical energy would have been transferred to smaller transmitters that would be in turn transmitted to local receivers on buildings, cars, boats, planes, etc. All electrical needs would have been powered wirelessly similar to the grid system employed by the Ancients.

According to Toby Grotz of the International Tesla Society, a wireless transmission of electricity has been proven. The conduit is the Schumann cavity, an area of atmosphere from the surface of the earth up to 80 kilometers. Experiments have shown that electromagnetic waves of extremely low frequency in the range of 8 Hz propagate with little attenuation around the planet within this cavity. Nikola Tesla in 1899 noticed the existence of the standing waves in this cavity normally generated by lighting. Tesla stated that these stationary waves "… can be produced with an oscillator." A working Tesla wireless power system would follow the surface of the Earth in all directions expanding and contracting until meeting a point opposite the transmitter where they would be reinforced and sent out again (just like the waves generated from lightning bolts). Telsa proved the existence of the waves in an experiment in Colorado Springs.[17]

Tesla began construction of his wireless power transmitter in Shoreham, New York, on Long Island. In the early 1900s it was dismantled before its completion. Tesla's inventions were far beyond his time and his wireless electricity theories posed a threat to the established AC/DC electrical industry. Therefore, further funding for the project was not forthcoming and we were left with strings of wires over countless billions of earth miles. But, a financially lucrative industry was preserved.

As proof of the current existence of wireless power, a photograph appeared in the May, 2004, edition of *The Smithsonian*

---

[17] Tesla, Nikola, *Colorado Springs Notes*, 1899-1900, Aleksandar Marincic, Nikola Tesla Museum, 1978

magazine[18] in which large fluorescent lights were inserted into the ground beneath an overhead high tension power line. As the sun set these lights became visibly lit – without being plugged into anything except the earth! Absolute proof that electricity can be transmitted without wires.

The Bermuda Triangle – Crystal energy in the now times

In the Bermuda Triangle, on the ocean bottom where the ruins of Atlantis are reported to exist, the energy built up in the sunken and damaged Fire Crystals could periodically trigger dematerializations of anything in the area. Electromagnetic disturbances are quite common in that area causing instruments to malfunction on boats and aircraft traversing the region.

In his *The Bermuda Triangle*,[19] author Charles Berlitz noted that many unexplained recorded events have occurred in this area. Apparitions have appeared to seamen. Ghost ships and planes have been seen. A Cessna 172 flew over Turks Island in the Bahamas on a clear day, and although people on the ground could see the plane and the controller talked to the pilot, no one in the aircraft could see anything but an undeveloped island. The plane was never found. Did it jump dimensions? There are numerous reports of dense fog banks in the area gobbling up vessels never to return, planes flying into clouds and never coming out the other side. If a pyramid or satellite transfer obelisk had sunk in the area of the Bermuda Triangle, could the streaming ions that are generated when the crystal unit is disturbed (such as in a plate shift) cause the magnetic interference that affect navigation instruments? Could this magnetic shift cause a dimensional jump or a dematerialization of existing matter as we saw in the Philadelphia Experiment during the 1940s?

This experiment was labeled "Project Rainbow" and was created to test Einstein's Unified Field Theory. The idea was to

---

[18] "Just Looking," photography by Lewis Whyld for South West News Service, Smithsonian magazine, May 2004, p28-29.
[19] Berlitz, Charles, *The Bermuda Triangle*, 1974, Doubleday, NY

make a ship invisible to the enemy in response to the vast sinking of allied ships by Nazi U-boats. It focused on the method of degaussing, a process in which a system of electrical cables are installed around the circumference of a ships hull, running from bow to stern on both sides. A measured electrical current is passed through these cables to cancel out the ship's magnetic field. This in effect made the ship invisible to radar returns. During the test, the ship (the U.S.S. Eldridge) momentarily dematerialized, and when it reappeared many members of its crew rematerialized dead inside the ship's bulkheads. Some seamen went mad due to the time shift experience. Officially, the Government stamped this project a failure and closed it. Rumors are that the technology underwent further testing at the Brookhaven National Laboratory following World War II.[20] Unexplained mysteries abound in the new millennium. Maybe we have to look to the Ancients for answers.

    The global power grid from the Ancients seemed to be a free-energy system. Our current society based on economical survival and greed has no room for free anything. So we string wires, erect nuclear reactors and charge lots of dollars to keep the lights on.

    Whenever inventive folks emerge with alternative power sources, the electric companies spend inordinate amounts of money and energy to put an end to that type of entrepreneurial spirit. This may not be a bad thing as the misuse of the free-energy power grid set up by the ancients could have played a part in creating the last great cataclysm that destroyed much of the planet including Atlantis. Power grid frequency may have been tuned too high for the purpose of increasing its destructive force during time of war. This could have been enough to activate seismic activity and set off volcanoes. Add this mayhem to a lively atomic war 12,000 years ago and you have the stuff the apocalypse myths are made from.

---

[20] Gordon, Wade, *The Brookhaven Connection*, 2002, Sky Books, New York

Aircraft

Then: According to Ruth Montgomery's channeled information, the Atlanteans discovered [that the] broadcast power of the Great Crystal could be used to propel vehicles. The Great Crystal was powerful enough that when directed through the rays of the sun it created sparks that were so strong ships would take off from the earth and move in currents directed by facets of the Crystal. Other vehicles powered by the same source swam beneath the seas. In fact, the Great Crystal could drive machines across the heavens and beneath the sea practically anywhere in the world.

Advanced vehicles were ultimately used for war. The Atlanteans and the Rama from India both had aerial capability and it was thought they even flew extraterrestrial with stories of lunar battles filtering down from the ancient texts. They refer to the airships as Astras and via them the ancient Indians could have sent a detachment of men onto any planet. These texts also referred to a cloaking device that made the ships invisible. The Atlantean and Rama societies were far superior in aviation technology. The Atlanteans being of a war-like temperament, created aircraft to levy high-tech weapons against the civilizations of the world. First created 20,000 years ago, these "fighters" were saucer or cigar shaped, similar in design to reported UFOs of the recent century.

In the ancient text Yantra Sarvasva written by the sage Maharishi Bhardwaj, strict construction specifications are laid out for military aircraft. They had to be impregnable, unbreakable, non-combustive, have hovering capabilities, invisibility, have eavesdropping mechanisms for identifying hostile aircraft, method for viewing inside aircraft, able to render the enemy crew incapacitated through suspended animation, be constructed from heat absorbing metals, and provide mechanisms that could enlarge or diminish images and sounds. Could these texts have inspired the writers of modern science fiction to create our TV starships?

Now: The new millennium civilization is no stranger to air travel. We have seen man's first attempts at flight transform into a flourishing industry. Behind most of the impetus for aircraft were the World Wars and once these battles were over, the nations turned to creating an industry that saw the expansion of aviation encompass airline, corporate and recreational flying. Just as the Ancients used air travel for business and pleasure, so have we made it a necessary part of life.

Behind most successful large aircraft manufacturers are military contracts that create momentum for aerospace advancement. New developments in aerospace technology are forged from a desire for military superiority. Area 51 in the southwestern U.S. has been a not-so-secret testing ground for many fighters, bombers and spy planes that eventually are revealed to the public eye. The Stealth Bomber is a fine example of an aircraft that meets many of the specifications laid out by the ancient texts. The composite material used on the Lockheed F117-A Nighthawk stealth fighter is lighter than aluminum and so strong it won't dent. Even the invisibility requirement has been satisfied. The successor to the Stealth Fighter is the Black Manta or Baby B-2, a flying wing that attains speeds far greater than the Stealth.

The Brilliant Buzzard or Mothership is thought to have a wingspan of 200 feet with the ability to launch smaller recon-naissance craft from within its hold. The Bird of Prey is a tactical fighter first flown in 1996 that uses new technologies enabling it to fly stealthily even during daylight. The recently revived Mach 3 SR-71 flies at 90,000 feet and has the ability to outrun SAM missiles. It is built of titanium and can sustain flight temperatures of 3,000 degrees. Many of these aircraft could be considered UFOs as they have delta shapes or look like designs from the Klingons. Current developmental aircraft reportedly fly at speeds that see travel from Arizona to New York in less than fifteen minutes.

Saucer-shaped craft have been reported near Area 51 as well as throughout the globe since the early part of the last century.

Could governments have recreated the technology so meticulously laid out in the ancient texts? Will we finally discover that UFOs are not alien structures but recreations of earthly vehicles from 20,000 years ago?

And, just like the Atlanteans, we are striving for more deadly aircraft to carry more deadly weapons. Flying machines were described in the ancient texts and now we have flying machines. Lasers were described in the ancient text and now we have lasers. Atom bombs were described in the ancient text and now we have atom bombs. Electromagnetic weapons were described in the ancient texts and now we have electromagnetic weapons. Are we good students or what?

Society and War:

Then: Several factions of thought and belief dominated ancient society. The early Atlanteans were a peaceful people. They recognized laws dictating that energetic forces governing the earth and the beings of the earth also govern the galaxy and the heavens. They cherished spirit and believed in a creative force manifested from the Universal Consciousness. As time progressed the Atlanteans matured further, and through the use of crystal energies were able to rejuvenate the human form and extend their lifespan by hundreds of years. Scientists of the Ancients created many beneficial procedures for curing ailments including laser surgery and light therapy. Ancient healers used the ley lines from the Earth's natural grid system to enhance health. They created ultrasound and used mind power by creating tonal vibrations; much like mantras employed in meditation today. Crystals were used to balance harmonies in the body to stabilize the flow of prana and to stimulate the chakras (key energy points in the body). Crystals of various colors were used to heal some illnesses.

At the same time other civilizations were developing along these lines. One in particular achieved the technological expertise of the Atlanteans. This society was centered in India and northern

Pakistan and referred to as the Rama Empire. Started by the Nagas who had come into India from Burma 15,000 years ago, they were a nation made up of large, sophisticated cities, many of which have begun to be found by archeologists. They were ruled by the benevolent aristocracy "enlightened Priest-Kings or Masters" who governed the cities. These were men possessed a high level of psychic abilities that they would seem godlike to people today.

The seven greatest capital cities of Rama as referred to in the Hindu texts, were "The Seven Rishi Cities." According to the texts the level of technology was equal to the Atlanteans complete with flying machines and electricity. At Mohenjodaro archeologists have found evidence of a well-planned city laid out on a grid with a plumbing system superior to those used in Pakistan and India today. The Rama cherished peace and harmony, prized spirituality and lived a more utopian life even though they continued to amass technological inventions of war.

The Atlanteans did not follow in the Rama footsteps. Their pastoral existence was replaced by one whereby greed and a thirst for materialistic comforts dominated their actions. This patriarchal civilization with their technological superiority deemed themselves "Masters of the World." Their goal was to take over other societies and to that end they developed highly sophisticated weapons. Brainwashing was not foreign to our ancestors. The scientists learned how to block brain impulses to parts of the brain that caused crime and negative emotions. Using crystals the Atlanteans projected detailed visuals into an unsuspecting person's brain. By doing this human thoughts could be influenced, as could memory banks. As the "dark priesthood" gained control of Atlantis in its latter stages, this form of "hypnosis" was used to manipulate the population. Misuse of crystal power was also used to create pestilence and diseases to kill people by holographically projecting the images, fears, concepts they wanted to impress on people. Just as today's philosophies claim that thoughts have the power to manifest reality, the Atlanteans used this power of the mind to their advantage.

Scientists experimented on the population by manipulating embryos to create subhuman forms to be used as slaves. Embryo development was arrested with crystal energy and hypnotic suggestion caused the embryos to stay more reptilians. Historians theorize that these lower forms of life were our ancestral cave men that materialized after the apocalypse. Eventually telepathic warfare, human sacrifice and a war of minds became rampant throughout the empire. Much of the population could see the downfall of civilization through misuse of these powers. They warned that Atlantis was poised to not only destroy their enemies, but itself as well. They voiced concern that the irresponsible use of technology could upset the planetary harmonics generating earthquakes and tidal waves. No one in power listened.

The Atlanteans were intent on conquering the Rama Empire. According to the Lemurian Fellowship the Atlanteans positioned their airships and armies outside a Rama city. The Priest King when asked to surrender said that the people of India have no quarrel with Atlantis. We ask only that we be permitted to follow our own way of life." The Atlanteans assumed the Rama had inferior firepower and began to march on the city. It is told the Rama used some sort of mind control tool causing the soldiers to drop in their tracks and retreat.

Not to be outdone in the war arena, the Atlanteans decided to unleash their most destructive weapon. Evidence points to the destruction of the Rama Empire through nuclear attack. The Rishi City of Mohenjodaro was excavated in the last century and gave all indications of falling during an atomic war. Once satisfied the Rama were extinct the Atlanteans continued on to the Gobi Desert to conquer the civilization there. The weapon of choice was a Scalar Wave (described as a present-day "laser-fusion reactor/gravity wave generator" by NASA researcher John H. Sutton) which was fired through the center of the earth. According to Edgar Cayce, intense gravity waves were generated by the Crystal and beamed into the Earth's crystalline quartz, which occurs in granite rock. This quartz crust absorbed the energy and the resulting meltdown of large masses of subterranean quartz would have been the triggering mechanism for destruction of the Gobi land. But the Atlanteans didn't figure on global repercussions. This wave also caused major slippage along the Earth's fault lines precipitating a global polar shift. In an effort to further utilize their Weapons of Mass Destruction other analysts theorize that the Atlantean's Great Crystals were actually tuned too high in an effort to escalate a war advantage. These energies were directed into the earth as well overloading the grid system and setting off harmonic waves that disturbed the tectonic plates, volcanic magma and the magnetic poles.

The legends of Atlantis sinking have been around for eons. Other theories locate Atlantis under the Antarctic continent – a civilization instantly frozen when the poles shifted. If the ancient texts and myths are accurate we can hypothesize that the technological warfare employed by the Atlanteans became the catalyst for the apocalypse. Sunspots are theorized to have affected the stability of the planet. During the time of the pole shift, magnetic pull from planetary alignments in the segment of the Milky Way the solar system was traveling through could have put stress on the planet's rotation. Could it be that these astronomical factors weakened the position of the earth so man's destructive forces could easily trigger the upset? Would the shift have occurred without man's intervention? Many scientists are seeing a similarity in the astronomical conditions that existed 10,000 years ago and those that are forecast for the year 2012. If we make similar technological mistakes or launch WMDs during this time period, could we trigger another apocalypse?

Now: It is interesting to look at the history of India post-Rama. According to Indian history, the civilization that flourished 4,000 years ago in the Indus Valley had planned cities with brick homes, drinking wells and private baths. Artifacts recovered reveal that children played with toys and women painted their lips. This information was discovered when, in 1922, archeologists found the remains of the ancient cities of Harappa and Mohenjodaro. Their time period was concurrent with the Egyptian and Sumerian civilizations. This civilization was highly advanced. Could they have used some of the technology that survived the Rama destruction?

As the floods common in the region destroyed the cities subsequent settlers rebuilt over the old foundations. Each city built on the old city became less and less sophisticated. Each generation seemed to be losing the knowledge of the Ancients. Somewhere around 1700 BC, a tribal nomadic peoples living in the far reaches of Euro-Asia began moving into Persia and India. Although not technologically advanced, these peoples were fierce and warlike, and gave themselves the name of "superior or noble." The name Aryan is derived from the Sanskrit work Aryas. It appears they may have been descendents of Atlantis having passed down the belief systems of that dominant society.

The Aryans (or Vedic) swept into India and dominated their civilization. They were a warlike culture that functioned in individual tribal, kinship units, the Jana that was ruled over by a warchief. Their society became very territorial each developing its own customs and languages. This is still evident today in India. The history of these peoples is called the Rigvedic Period (1700-1000 BC). They originally only had two social classes: nobles and commoners. Eventually they added a third: Dasas (the dark skinned peoples they had conquered). Finally the cultural mixing with the indigenous cultures reduced the dominance of the war-religion of the Aryans. By 200 BC, the process of a more benign transformation was more or less complete and the culture we today call "Indian" was fully formed.

The warlike Atlantean heritage seems to have dominated Europe and thus North America. If we look at history, we can see a repetitive pattern of war and conquest that continues to this day. The most powerful countries seem to carry the biggest stick. People all over the globe have difficulty living in peace. Could this propensity for global domination have been passed down in the genetic codes established by the Ancients? If we make an analogy to the ultimate downfall of the Atlanteans we can see a similar pattern emerging today.

As the Atlanteans drifted from their spiritually harmonious lifestyle into greed and dominance, their regard for Mother Earth vanished. Creature comforts ruled their society at all costs. Are we following this same pathway? Current technology has accelerated at a mind-boggling pace since the late 1800s. Baby boomers remember when TV first emerged. Their parents can remember the launch of the automobile and their grandparents saw the first electric light bulb. In the USA we are now a wireless society with communication tools designed to fit in a pocket and a throw-away society that ships waste to third world countries.

As a society, American leaders seem to be recreating the Atlanteans quest for power and greed along with mimicking their disregard for planetary health. Our environmentalists have lots of passionate causes to lobby for with an equally staunch opponent that touts convenience and fear of lack as justification for looking the other way. Precious rangelands are opened up for oil development at a cost of thousands of wild horses. Arctic refuges are being spoiled because of the greed for oil with indigenous animal species migratory pathways altered because of pipelines. Cell towers and powerlines are ruining the landscape at an accelerating pace with power and communication companies disregarding research on how their frequencies may affect humans and the planet. Our throw-away society is leaving a legacy of plastic and heavy metals that may be excavated by archeologists in the future. Disregard for the creation of sustainable energy sources is increasing our carbon footprint. Yet we have progressed to a convenient lifestyle that, in the USA, employs the latest technological marvels in a world where across the globe whole populations have no TV, computers, cars, indoor plumbing or cell phones. It would seem that the USA is repeating Atlantean history.

The European Union seems to be evolving from their Atlantean heritage and employing many more of the Rama values. They have established a higher quality of life by implementing many environmentally friendly programs that will tend to make Europe a flower within the dark, smoky industrial world. The EU is coming around to noticing that man is not alone on the planet. They are structuring new programs that integrate modern life with a harmonious interaction with the environment and their fellow man.

As noted by Jeremy Rifkin in the March/April, 2005 issue of *E Magazine*,[21] the USA and the EU have opposite approaches to the question of environmental stewardship. Europeans are more mindful of the dark side of science and technology and can see the long-term destructive effects. In recent years the EU has boycotted genetically engineered food and refuses to allow US imports of genetically modified foods. They are stressing the importance of organic farming and have proposed new regulatory controls on chemicals that add toxins to the environment. This is diametrically opposed to the way Americans view the use of chemicals. The EU also has adopted a rule which prohibits electronics manufacturers from selling product in the EU that contain mercury, lead and other heavy metals. The USA is not so conscientious and in fact exports computer waste to Third World countries where street merchants melt down the parts to extract the heavy metals, a process that affects the health of the workers.

The EU recently adopted a new policy to regulate the science behind new technologies. The potentially dangerous effects of each proposed "invention" is considered as it pertains to environmental and health impacts. If there is any uncertainty regarding the proposed scientific venture, they can suspend the activity. This policy has found its way into many treaties including the Stockholm Convention on Persistent Organic Pollutants (2001).[22] This shows a willingness to embrace sustainable development, a term that elicits much resistance in the USA.

---

[21] Rifkin, Jeremy, "The European Dream," E Magazine, March/April 2005, p34
[22] Rifkin, Jeremy, *The European Dream: How Europe's Vision of the Future is Quietly Eclipsing the American Dream*, Tarcher/Penguin, 2004

In correlating the viewpoints of the leaders from the USA, we find a striking similarity to the blindness the Atlantean leaders exhibited as they grew closer to fatally impacting the planet. Americans are divided in their support of governmental policies and many warn of impending destruction if we follow our current insensitive pathway. This is repetitive of the warning from the Atlantean peoples to their leaders, which fell on equally deaf ears.

Our European counterparts have voiced concern over many policies dictated by the USA. New "weapons" have surfaced and the EU is asking for further research prior to releasing them. One of these is a non-lethal crowd control device. The EU is concerned that not enough testing has been done to determine the biomedical effects on the crowd. They have asked that research done on these devices be made available to the EU for review prior to implementation. These devices include plastic bullets, pepper gas and microwave bursts that burn the skin. The USA currently employs these types of devices (the latter deployed in Iraq in 2005), with most of the research having been done in secrecy.

Little notice has been taken on the professional hazard assessments of the most commonly used kinetic impact weapons that have consequences in the dangerous or severe damage region. These include ultrasound generators which cause disorientation, vomiting and involuntary defecation. Visual stimulus and illusion techniques such as high intensity strobes can cause epileptic seizures. A disabling, calmative, sleep inducing agent mixed with DMSO when released crosses the skin barrier quickly causing pain, paralysis and general disablement that could be lethal to sensitive individuals. They also have foam guns that stick the targets feet to the pavement, human capture nets laced with chemical irritants, blinding laser weapons, thermal guns that raise the body temperature to 107 degrees and a magnetosphere gun which delivers what feels like a blow to the head.[23]

Other non-lethal weapons that mimic those used by the Atlanteans are being introduced worldwide and their production

---

[23] Begich, Nick and Roderick, James, *Earth Rising II*, Earthpulse Press, 2003

is becoming a booming industry. Surveillance and identification equipment is now widely accepted thanks to the events of September 11, 2001. Biometrics that read the iris, fingerprints, DNA, odor and signatures are finding their way even into laptop activation programs. At first these seem harmless until one realizes the database that must be created to sustain such an operation. Where is our right to individuality and privacy? Our information is now available to anyone with access, including those who target political activists or insurgents for elimination. Using data profiles, torturing states have used these systems to compile death lists. A priest finding himself on the list escaped from Guatemala saying, "They had printout lists at the borders and at the airport. Once you found yourself on the list it was like bounty hunters were after you."

Even the transponder scanners I described in *2012 Airborne Prophesy* for use by travelers are currently being used to track prisoners on probation. It is just a matter of time until these manufacturers will expand the marketplace beyond the prison system. A patented device (U.S. 5,629,678) is already in the works as an implant. Power for the remote-activated receiver is generated electromechanically through the movement of body muscle. We have already seen these in use for animal identification and recently a push is targeting children. The upside is that you can always locate the person. The downside is that you have lost all sense of privacy and "Big Brother" will always be watching you. Also in the works is a patent for a bar code tattoo that the wearer uses for access to public places, banking and sales transactions. Again – convenience or loss of freedom?

As if its not enough that man spends an inordinate amount of resources developing ways to watch and kill their fellow man, the new trend is to disable information so the "enemy" will be incapacitated and become an easy target. As more and more dependence on electronics and computers become an essential part of a country's defense system, new information weapons are surfacing as disrupters. Most of us have experienced a hacker's wrath introducing a worm or virus into our computers. This Infowar is now being expanded to include military defense computers, launch sequences, satellite disrupters, Internet sabotage. If we're not careful it will be a simple keyboard task to push someone else's red button and launch a missile. Great resources are being implemented to monitor the networks for hackers and those engaged in suspicious activity.

Although computers have accelerated society's information and communication channels it also has made us vulnerable to mind control, suggestive rhetoric and hacking. As the Atlanteans became masters in creating weapons of mind control, we too are faced with victimization. The electromagnetic spectrum carries waves that if tuned to the brain's frequency can impart subliminal messaging. The Internet can also have this effect as can radio and television. People seem to believe the media regardless of the validity of facts. Using electronic devices to change thought patterns, and the media to reinforce the "truism" of these "beliefs," resistance to governmental policies can be eliminated. What a better way to gain control of the world without using destructive weapons. Unfortunately, the Atlanteans let emotion mar their judgment and fired on the Rama before the electronic messaging had time to make the "enemy" subservient.

Our acceleration of technology has been compressed into far fewer years than it took the Atlanteans to emerge as a dominant power. We are not reinventing the wheel but merely digging up the past. How this information was made available to modern man is a mystery to most of us. I'm sure the core rulers must have archives of instructional manuals. It is said that the aliens that crashed at Roswell gave us technology that launched a new era. Were they aliens or had we discovered the ancient instructional manuals and built our own ship that crashed? The UFO sightings may be nothing more than these experimental aircraft recreated from the Atlantean/Rama era and flown by our own military. If we have reached a similar level of destructive power we may be poised for another cataclysm.

The USA seems to be the dominant aggressive power as were the Atlanteans. Wars that are created to fund, for example, phantom WMDs, only serve to expand our dominance for reasons not known to the general public. Other countries exhibit aggressive behavior but pale compared to the strength and power of the USA. September 11, 2001, served well to enforce our fears and gain support for the government's aggressive policies. Debate continues as to whether this was an act of violence against our nation or merely a staged event to sway the support of the American public. Regardless of the truth it was an effective means to rally the troops for unnecessary aggressions. Just as the Atlanteans discounted public opinion, many Americans also feel helpless to challenge government policies whether environmental or military. Our rulers seem to be reenacting history and following step by step into another Atlantean destructive cycle. The means to the end of the world may have changed. Atlanteans had crystal power. To our knowledge this is not being employed today, but we have energy sources that equal the volatility of crystal energy and can destroy the planet as completely.

Chapter IV:
Energy Waves Leading Us To Destruction

Nikola Tesla was an immigrant from Yugoslavia. Born in 1857, many contend he was an Atlantean engineer reincarnated. As the inventor of AC current, Tesla went on to construct marvels of electricity. He worked with radio-frequency electromagnetic waves and developed what we know as radar. His vision was well over a century ahead of his peers. Just before the turn of the century he began experiments with wireless electricity employing a magnifying transmitter capable of generating 300,000 watts of power. This is similar to the technology used in the High Frequency Active Auroral Research Program (HAARP) that currently poses a threat to the stability of the ionosphere.

    Tesla believed that energy could be pumped into and extracted from the earth at the nodes of terrestrial stationary waves (similar to the Ancients' generation of crystal power at certain grid points). He demonstrated this in one experiment where he lighted 200 of Edison's biggest lamps from a distance of twenty-five miles – without any wires. This was similar to the Atlantean power grid system without the crystals or lightning generator. So impressive was the potential of his broadcast power, J.P. Morgan agreed to finance the construction of a tower on Long Island, NY. Unfortunately, financial support was pulled prior to completion and the tower was dismantled. Tesla said this transmitter could have produced 100 million volts of pressure with currents up to 1,000 amperes, which is a power level of 100 billion watts. Such a transmitter would be capable of projecting by radio, the energy equivalent to a nuclear warhead. Any location in the world could be vaporized at the speed of light. It also could be used for wireless electricity and to power airships.

    One test of his transmitter is suspected in creating the devastation in Siberia on June 30, 1908. An explosion of staggering proportions took place near the Tunguska River in Siberia.

Estimates on the force of the explosion was exactly what Tesla's device could produce. Researchers have attributed the devastation to either a meteorite or a comet that exploded above the area wiping out trees, wildlife and two people. There were no fiery objects observed in the area, leaving the explanation of a Tesla experiment gone awry to gain more validity. There was a glow observed over the area, which is consistent with what would happen if a high current was directed through the earth to this area. Basically it is thought that Tesla was aiming for Alert, Canada, and overshot his mark. If this weapon in fact created the devastation, then it proves that Atlantean technology had reached modern man. A noteworthy aside is that, upon Tesla's death, a U.S. Government official confiscated his files. Could our modern weapons be an outgrowth of Tesla's inventions? The Russians took Tesla seriously and it is thought that their Woodpecker transmitter is an outgrowth of the Magnifying Transmitter. Although reportedly used to detect incoming aircraft, it was discovered that the Russian Woodpecker could transmit at a frequency that affected the nervous systems of humans and thus could be used for mental manipulation of target populations (including the United States Embassy in Moscow, Russia).

Dr. Andrew Michrowski, Ph.D. said of the Woodpecker, "Since October, 1976, the USSR has been emitting extremely low frequency signals from a number of Tesla-type transmitters. Their frequencies correspond to brain-wave rhythms of either the depressed or the irritable states of humans. The Soviets are on the verge of a breakthrough that will make missiles and bombers obsolete. They could disrupt cities by inducing panic or illness into whole nations just by sending out radio pulses."[24]

The actual date of the discovery of suspected transmissions from the Woodpecker was in 1953. And in 1992, the Washington Times reported that "the Russian Government is continuing to

---

[24] Smith, Jerry E., HAARP, *The Ultimate Weapon of the Conspiracy*, Adventures Unlimited, 1998

bombard the U.S. Embassy in Moscow with microwave radiation."[25] This process actually captures the brain waves of the people affected by putting them in sympathetic resonance with the signals. In this state the person wouldn't realize they are being affected and would exhibit behavior or emotional changes at the will of the person operating the transmitter. Victims claim they developed many symptoms including a rare blood disease, suffered from bleeding of the eyes, headaches and cancer. A 1990 article from *The Atlantic Monthly* warned that "the Soviets had the RF technology to build weapons that could degrade or destroy electronics or disorient personnel. If an RF pulse could be propagated over a wide zone it might act as an electronic wall disabling the silicon brains of any approaching airplane, tank or missile."[26]

This possesses a striking similarity to the Atlantean and Rama civilizations' mind control weapons. A closer look into brainwashing techniques from the character's perspective is expanded upon in *2012 Airborne Prophesy*. Sophisticated military weaponry was described in a 1993 Defense News article that claims acoustic psychocorrection involves the transmission of specific commands via static or white noise bands into the human subconscious without upsetting other intellectual functions.[27] Could our government be using this remote EEG technology to discredit the claims by war victims involved in a former intelligence operation? Could this device be used to alter the credibility of protestors, foreign heads of state, ambassadors and political opponents?

In 1973, Walter Reed Army Institute of Research discovered that externally induced auditory input could be achieved by means of pulsed microwave audiograms. The effect on the recipient is the schizophrenic sensation of "hearing voices"

---

[25] *Washington Times*, Nov. 15, 1992
[26] Smith, Jerry E., HAARP, *The Ultimate Weapon of the Conspiracy*, Adventures Unlimited, 1998
[27] "U.S. Explores Russian Mind-Control Technology," Defense News, Jan. 11-17, 1993

which are alien to the person's own thoughts. These weapons are used to temporarily drive a target crazy. High-tech electromagnetic pulse devices have been used to physically alter human impulses such as anger or libido. This weaponry has also been used to destroy electronic circuits, fry computers, disable automobile and truck engines and creates holographic projections to deceive an attacker. Unfortunately the safety of such devices is in question. When aiming at the engine of a car the weapon can also stop pacemakers installed in heart patients who happen to be within a block or two of the target.

Several very low frequency devices have been created to induce nausea, vomiting and abdominal pains as well as disrupt human organs and crumble masonry. The US Federal Government has acquired enough high tech psychotronic devices to zap the minds of targeted victims and even create the appearance of UFO abductions as a method to conceal programs that inject computer tracking chips into individuals. Because these frequency levels are so low the target is unaware that any changes have taken place.

These electromagnetic mind weapons pale in proliferation to what I consider the greatest assault on the pulse of the earth. The cellular industry has created an indispensable convenience, but the towers and phones are altering the earth's primary frequency. We have filled the previously empty electromagnetic spectrum between these two extremes with man-made radiation that never existed on earth. Life on the planet evolved under geomagnetic fields and electromagnetic forces that have been constant for hundreds of millions of years. Over 100 years ago that changed when the first power station was built in New York City. Before 1900, the earth's electromagnetic field was composed simply of the field and its associated micro pulsation, visible light and random discharges of lightning. That has all changed today.

Our electric power systems operate at 50-60 Hertz (Hz), or cycles per second, just above the highest naturally occurring frequency of 30 Hz. Microwave and low frequency radiation is all around us coming from television and radio towers, satellites, high-voltage electric lines, computers, radios, paging systems, alarm systems, telecommunications systems, signal generators, electronic games and microwave ovens. Microwave beams operate at billions of times per second and are getting ever closer to the trillion cycle frequencies of visible light. The exposure of living organisms to abnormal electromagnetic fields results in significant abnormalities in physiology and function. If the electrical human being is affected, imagine what Mother Earth is experiencing.

Personal desires for the proliferation of frequency devices has seen an explosion with the advent of microwaves, computers and cell phones. Microwaves secrete their hideous frequencies from ovens and other devices that potentially could create human and planetary anomalies. These short waves of electromagnetic energy travel at the speed of light. Microwaves cause polar molecules to rotate at a frequency of a million times per second. This agitation is what heats the food and burns the skin when microwaves are used as weapons. This friction also causes substantial damage to the surrounding molecules forcefully deforming them and neutralizing the electrical potential (the cells' life force). Eating microwaved food passes along these dysfunctional molecules to our bodies, reducing the electrical charge. Microwaves have been shown to create a breakdown in the human life-energy field so serious that the Russians set strict limits of ten microwatts exposure for workers and one microwatt for civilians. This limit is disregarded by U.S. manufacturers who allow Americans a greater exposure. Microwave sickness includes headache, dizziness, eye pain, sleeplessness, irritability, anxiety, stomach pain, nervous tension, inability to concentrate, hair loss and even cancer. Changes are observed in the blood chemistries and the rates of certain diseases among consumers of microwaved foods. We can measure effects of these waves on the human body, but how are they affecting the planet's pulse?

Cell towers and cell phones have saturated what little free frequency space we have left in the atmosphere. No matter where you are on the planet frequencies alien to the body are affecting you. Radiofrequency (RF) energy source is in its antenna. Since it doesn't create heat we have been led to believe that unlike microwave energy, it is not harmful. Even fairly low levels of electromagnetic radiation can change the human body's sleep rhythms, affect the body's cancer-fighting capacity by harming the immune system, and change the nature of the electrical and chemical signals communicating between the cells. Studies have documented cellular health effects primarily because it takes time to pinpoint each cause and effect and the cellular industry is still young.

    It is known that some of our biological electrical activities can be interfered with via oscillatory aspects of the incoming radiation. This radiation seems to affect a variety of brain functions including the neuroendocrine system. The brain reaches peak absorption in the UHF bands, right where cellular telecommunications operate. A study by Dr. Peter Franch found that cells are permanently damaged by cellular phone frequencies and the damage is inherited unchanged from generation to generation. There are sixty-six epidemiological studies showing that electromagnetic radiation across the spectrum increases brain tumors in humans. Two of those studies are related to brain tumors from cell phones. The fear of cancer aside, cell phones are implicated in reducing the protection of the blood-brain barrier which keeps out toxins. It also as been suggested by University of Washington researcher, Henry Lai, that these frequencies also inhibit the uptake of calcium by the brains' neurotransmitters precipitating the long-term possibility of memory loss. Whether you believe these studies or not, cell phone and cell tower frequencies are alien to our bodies. Since we pulse at a rate similar to the earth we can also imagine how severely we are disturbing the harmonics of the planet.

    Further applications of electromagnetic pulses have been surfacing in recent decades. Since July, 1976, ten-pulse-per-second electromagnetic low frequency waves have been utilized for weather war attacks against America. These seem to be traced to the Russian Woodpecker, a transmission tower whose frequency emissions have been linked to applications for changing brain wave patterns. ELF waves at the frequency of the mind's alpha waves have been sent from the space shuttle. For what purpose – perhaps mind zapping? Dr. Gordon J.F. MacDonald, President Johnson's former science adviser, commented that these ELFs were "to create within the cavity between the electrically charged ionosphere in the higher part of the atmosphere and conducting layers of the surface of the earth, electrical waves that would be tuned to the brainwaves. The natural electrical rhythm of most mammalian brains is 10 Hz (or cycles per second), and there are indications that if you tune in at this frequency you can produce changes in behavioral patterns or responses." This was divulged in 1972. How far have we come since then and how active have these attempts at global brainwashing been, not to mention how frequency saturation may affect the weather and the planet's vibration?

As mentioned before, our electric power systems operate at 50-60 cycles per second (Hz). The naturally occurring pulse of the earth is 30 Hz. Microwave beams operate at billions of times per second and are getting closer to the trillion-cycle frequencies of visible light. The empty electromagnetic spectrum is quickly filling up between these two extremes with man-made frequencies that haven't previously existed on earth. Even the ancients with their power grids didn't seem to have the wide array of frequencies generated today. Because we are surrounded by these EM energies, the human body is being affected all over the planet. We are living, full time, in radio-frequency plasma.

As if it's not enough that we are saturating the lower atmosphere with frequencies alien to the planet, we are now targeting the ionosphere. A diagnostic system used to characterize the structure and dynamics of small-scale irregularities in the ionosphere (ionized portion of the layer between 35 and 500 miles above the earth's surface) is a VHF Satellite Scintillation monitor known as HAARP.²⁸ Scientifically useful irregularities can be produced by an ionospheric interaction facility like HAARP when its transmitted HF radio energy is converted to heat in the ionosphere over the facility, weakly perturbing the distribution of the background ionization.²⁹ In plain English, the heating of plumes in the ionosphere causes that segment to extend in height, creating a dome or bubble. This is useful as a mirror that can bounce radio waves off the ionosphere to a designated location. Without this ionospheric heater, radio waves bounce indiscriminately off the ragged ionosphere creating uncertainty as to where they will be received. This type of device can improve communications and control its trajectory with global positioning accuracy.

As benign as it sounds, this new ionospheric heater is far more potent (there are many already in various locations on the planet). HAARP employs a high power transmitter and antenna array operating in the High Frequency range. It can produce up to 3.6 million Watts to an antenna system consisting of 180 crossed dipole antennas arranged as a rectangular, planar array. It is capable of beaming in the 2.5-10 megahertz frequency range at a level of more than three gigawatts of power. Remember that the 10 Hz frequency is synergistic to the human brain's frequency. Could an underlying purpose of HAARP be for global brainwashing? In *2012 Airborne Prophesy,* HAARP is being impersonated by DRUM (Direct Radiofrequency Umbrella) and is implicated in a scheme to influence the thoughts of human targets.

The highly charged particles in the ionosphere could selectively cause interference with or even total disruption of communications over a large portion of the earth. They also could disrupt the guidance systems in airplanes and missiles, act as an eavesdropping intelligence tool and cause missiles or aircraft to fall out of trajectory because atmospheric heating changes the air density and coefficient of lift on the wings.

---

[28] *Satellite Scintillation Monitor*, http://w3.nrl.navy.mil/projects/haarp/satsin.html
[29] Patent # 4,686,605 held by Arco Power Technologies, Inc. (contracted to build HAARP)

In my novel the jet aircraft flown by the heroine actually finds this weapon useful to dispose of incoming enemy "bogies." Bernard J. Eastlund's US Patent #4.686.605 says: "This invention provides the ability to put unprecedented amounts of power in the Earth's atmosphere at strategic locations and to maintain the power injection level, particularly if random pulsing is employed, in a manner far more precise and better controlled than heretofore accomplished by the prior art, particularly by detonation of nuclear devices of various yields at various altitudes." It goes on to state: "…it is possible not only to interfere with third party communications but to take advantage of one or more such beams to carry out a communications network even though the rest of the world's communications are disrupted … large regions of the atmosphere could be lifted to an unexpectedly high altitude so that missiles encounter unexpected and unplanned drag forces with resultant destruction."

During active ionospheric research, the signal generated by the transmitter system is delivered to the antenna array, transmitted in an upward direction, and is partially absorbed at an altitude between 100-250 km, in a small volume a few hundred meters thick and a few tens of kilometers in diameter over the site. This technology will not only support communications and surveillance but will enable the military to send signals to nuclear submarines and to peer deep underground. Electromagnetic waves sent subterranean bode an eerie similarity to the technology of the Ancients gone awry. Some opponents "to this EHF weapon" have dubbed it the "Pentagon's doomsday death ray."

The patent that HAARP seems to be based upon, states that weather modification is possible by altering upper atmospheric wind patterns or altering solar absorption patterns by constructing one or more plumes of particles that will act as a lens or focusing device.

It states, "Weather modification is possible by, for example, altering upper atmosphere wind patterns by constructing one or more plumes of atmospheric particles which will act as a lens or focusing device ... molecular modifications of the atmosphere can take place so that positive environmental effects can be achieved. Besides actually changing the molecular composition of an atmospheric region, a particular molecule or molecules can be chosen for increased presence. For example, ozone, nitrogen, etc. concentrations in the atmosphere could be artificially increased."

Opponents of HAARP have suggested that the severe weather in the southwestern and central USA has been triggered by the operation of the ionospheric heater aimed over the western US. Since I have been flying as a commercial pilot for over thirty years, I have witnessed a profound change in severe weather. In the 1970-80s, it was rare to see a tornado in the Northeast or in Southern California. They are far less rare today. Temperature extremes are becoming the norm with record highs and lows being experienced in the same week. Thunderstorm activity seems to have increased with individual cells being far more lethal than twenty years ago. I can only assume that as we generate more and more alien frequencies, we add to the instability of the atmosphere and generate weather anomalies including devastating hurricane seasons such as the 2004 battering in Florida.

The reason Alaska was chosen as the location for the HAARP array is because the magnetic field lines, which extend to desirable altitudes for this invention, intersect the earth in Alaska. Another consideration is HAARP is located in an area that has ample access to natural gas as its fuel source. The HAARP technology has been equated to an invention created by Tesla who said, "my teleforce would melt airplane motors at a distance of 250 miles so that an invisible "Chinese Wall of Defense" would be built. This force would operate through a beam one hundred-millionth of a square centimeter in diameter." The similarity of Tesla's inventions to the HAARP technology is remarkable. Could Tesla, as a reincarnated Atlantean, have modernized the crystal beam that transmitted communication, power and a death ray?

It is suspected that this modern day "death ray" could actually be the signal sent out by HAARP to create a hole in the ionosphere allowing deadly gamma radiation to fall over a specified area. These gamma rays are high-energy radiation that can cause cell mutations and penetrate even thick concrete walls. As a weapon this is a fail-safe device that can be targeted from a remote location. This weapon would be the ultimate silent but deadly device. HAARPs other "death ray" is actually sent underground and used as "radar" to identify military installations. It is difficult to comprehend that an intelligent being would purposely send an alien frequency into the earth without consideration for triggering earthquakes or instigating volcanic activity – or maybe that is the purpose. A weapon that appears like a natural disaster!

We are not satisfied with atmospheric and ionospheric manipulation, but have set our sights on the electrojet as well. The electrojet is what causes the Northern and Southern lights (the aurora). To explain what the electrojet is we must examine the earth as a giant magnet surrounded by invisible lines of force. Incoming cosmic and solar radiation is trapped between these magnetic lines of force in the Van Allen Radiation Belt. Electrons also spiral around those magnetic lines of force with most of them bouncing back before entering the atmosphere. This bouncing back and forth creates a flow of electric current of great power. Some of this flow escapes and streams down on the earth at the poles. This stream of electrons falling to earth is called the electrojet. As they enter they heat the gases in the atmosphere and they give off photons (light) which is what causes the aurora lights. Since these electrons create 0.1 to one million megawatts of power, the HAARP technology is poised to tap into this energy source turning the electrojet into an antenna for communication. Operators plan to resonate their frequencies to match the vibrations of the electrojet greatly enhancing their rebroadcast potential especially targeting communications channels of vessels beneath the sea. This technology could be expanded to deliver the 10 Hz frequency message to the minds of unsuspecting targets.[30] Regardless of our fears of losing the choice

---

[30] Smith, Jerry E., *HAARP The Ultimate Weapon of the Conspiracy*, Adventures Unlimited Press, 1998

of our thoughts, no one has considered the environmental effects on the electrojet once we start manipulating its pulse. Mother Nature is being tickled and she may start twitching throwing weather and seismic curve balls at us.

Using radio-frequency technology for weather modification prompted the United Nations to draw up the 1977 Environmental Modification Techniques Convention (ENMOD). Many nations including the USA signed the "law" which prohibited the use of deliberate manipulation of natural processes to cause such phenomena as earthquakes, tidal waves, and changes in climate and weather patterns. Obviously, if the UN thought it important enough to create a mandate, there must have been technology in place to alter the weather as a means of warfare. The use of radiofrequencies makes it much easier than using cloud seeding to disturb the atmosphere. By electronic modification you can ionize or deionize specific segments of the atmosphere changing weather over exact locations.

In 1997 a Russian firm offered to use man-made "cyclones" to clear smoggy skies of Malaysia – an obvious admission to weather manipulation. VLF frequencies from power lines have been shown to affect the ionosphere and in reverse if we artificially influence the earth's aurora with small amounts of energy, we see a change in the outburst of the northern lights.[31] If it's true that the ionospheric heaters create a dome of higher temperature air, that dome could also serve as a lens, condensing solar energy to specific points below creating temperature changes that precipitate wind patterns. All weather is generated by instability in the heating/cooling of air that generates a pressure gradient and wind. If a device can change the temperature of the air it can change the weather. Some scientists warn that fooling around with the weather will cause irreversible damage to the atmosphere.

Weather manipulation is not the only threat to our planet's stability. Techniques to generate earthquakes pose more immediate and a devastating danger. In 2004, there was an earthquake

---

[31] de Caro, Chuck, "The Zap Gap," *The Atlantic*, March 1987, p 24-28

in western Colorado that was linked to a man-induced military operation. Could the Tsunami that hit the Indian Ocean in 2004 have been caused by a man-made earthquake? Just like it is suspected that Tesla missed his mark and leveled part of Siberia, maybe a subterranean radio-frequency experiment missed its mark, too. Tesla did generate an earthquake in New York City causing substantial damage and frightening countless residents. The *New York American* in 1935 linked Tesla's experiments as transmitting mechanical vibrations through the earth and causing a controlled earthquake. In an interview with *World Today*, February, 1912, Tesla claimed to have the technology to split the planet by combining generated vibrations with the earth's own vibration. Similar to the death ray supposedly used by the Atlanteans, Tesla's invention has been linked to events in and around his laboratory in New York City and the Siberian mystery discussed earlier in this book. Since his design papers were confiscated by the government after his death, we can only guess that *somewhere* his earthquake machine is alive and well and poised to be used as a "natural" weapon. Can cataclysmic events such as earthquakes and hurricanes be triggered to shift media focus from unpopular political agendas during critical times such as election years?

As the Atlantean power figures ignored human suffering and dignity in favor of dominance and greed, so too may our power figures forget that they come from the same species as those they are trying to dominate. We have created so many weapons of mass destruction, from biological warfare to weapons that injure innocents, kill people and change the minds of the "enemy" – and, as a sidebar, can destroy the very planet we exist on. It is not unthinkable that we have created technology that can alter the weather and create natural disasters – Tesla had the power a century ago. We have come far from the creation of the light bulb and it is plausible to assume the Tesla-Atlantean technology is currently being used.

Chapter V:
Accelerating Mother Nature's Rage

In the 19th century there were only 2,119 earthquakes recorded. In 1970 alone, there were 4,139. The largest earthquakes on record have all occurred since 1960. Volcanoes too have been erupting in force. At the beginning of the twentieth century, few volcanoes were active. By 1960 an increase in activity has been reported, and in 2004 eight volcanoes spewed their insides beginning in July with more becoming active thereafter (ten became active on October 5, 2004). Tornadoes in 1950 numbered 4,796, but in the 1990s they were in excess of 10,000. The percentage of the population affected by natural disasters tripled since 1900 to 2003.[32] Earthquakes are increasing with the most visible being the plate shift that caused the devastating tsunami in the Indian Ocean in December, 2004. Hurricanes are on the increase. Not only in numbers but also in severity, and as we saw in Florida in 2004 – in repetitive sequencing. Is this just planetary evolution or is man the trigger?

Astronomical occurrences affect the stability of Earth and are known to affect the cycles she goes through. Interplanetary magnetic fields (IMF) are weak magnetic fields that fill interplanetary space with field lines usually connected to the Sun. IMFs play a part in the flow of energy from the solar wind to the Earth's environment. In the absence of substorms from the Sun, the IMF conditions enhance the electrojet with a flow of activity without energetic particle injections. These IMFs sometimes contact the Earth's magnetosphere. The spinning liquid center core of the planet creates a dipole magnetic field, which is distorted by the energy of the sun into what is known as a magnetosphere. This field changes as the earth rotates toward

---

[32] Lockhart, Gary, *The Weather Companion*, Wiley and Sons

and away from the sun creating daily fluctuations in strength. It is what causes biological rhythms in human beings.

The magnetosphere is being affected by the planetary wobble and, along with the changing magnetic pole alignment, results in a feeling of time speeding up. Many, many people are complaining today that they just feel the days are too short to get everything done. The magnetosphere can be disturbed by magnetic storms generated by sunspots and Coronal mass ejections (CMEs) which generate ions that we sometimes see as an aurora. Sunspots are blobs of plasma that the sun spits out as nets of charged particles traveling at monumental speeds through space. When these hit earth, the charge covers the surface creating a "short circuit" that if intense enough (such as in a CME) can slow the Earth's rotation or shift its crust. These CMEs could alter the weather and possibly shift magnetic north.

Sunspot activity at the time of the Earth changes during the Atlantean age could have weakened the planetary stability enough to allow the man-created power grids to further upset Earth's balance creating a polar shift. We do know that there was a rapid temperature instability on the planet 10,000 years ago initiated by a chain of events from highly charged material in interstellar space breaking into our solar system. A similar trend is seen today. This high-energy atmospheric phenomena is now becoming more intense revealing itself as linked to the number, frequency and magnitude of natural disasters from earthquakes and volcanic eruptions to severe storms. Mayan predictions about the potential for planetary disasters in 2012 took into consideration their plotting of future major CMEs.

The ionosphere is a layer of atmosphere containing an appreciable number of ions and free electrons. Here the ions (atoms with an electrical charge) are created by sunlight (plasma generation). When the planet rotates away from the sun these ions and electrons recombine reducing the plasma in the ionosphere. The ionization is affected by sunspot cycles, seasons, latitude and solar-related ionospheric disturbances. It increases in the sunlit atmosphere and decreases on the shadowed side. The ionosphere is a dynamic system controlled by many parameters including acoustic motions of the atmosphere, electromagnetic emissions and variations in the geomagnetic field. In some circles it is thought that there is persuasive evidence of an ionospheric precursor to large earthquakes that can be used as a predictor. Prior to large earthquakes electromagnetic emissions (EMEs) have been detected in the ionosphere up to six days before the event. The ionopshere supplies plasma to and interacts strongly with the magnetosphere.

The electric currents that connect the ionosphere to the plasma sheet and the magnetospheric boundary layers produce an electric field across the polar cap which generates horizontal currents in that region of the ionosphere. The result is conductivity of the high-latitude ionosphere supplying a trillion watts of heat to the upper atmosphere, dramatically altering global thermospheric wind patterns during times of strong geomagnetic activity. The ionosphere is a key player in life on planet Earth. It absorbs harmful solar radiation, x-rays and ultraviolet rays. At its lower levels, it can reflect radio waves making possible transmission to far locations.

The atmosphere, ionosphere, magnetosphere and electrojet exist in a delicate balance that serves to make our planet habitable. Upsetting "the apple cart" may prove disastrous. Man notoriously blunders into areas of experimentation without exploring side effects. As we have seen of late with numerous drug recalls, how many mistakes can we write off with just an "oops?" When we are dealing with planetary harmonics, we may not get a second chance to correct our errors. Critics of the high powered ionospheric heaters claim that their gigawatt beams will not just burn a momentary hole in the ionosphere, but create a tear that could self-perpetuate allowing deadly radiation to stream to the surface unchecked. These frequencies may also alter the normal pulse of the geomagnetic grid that surrounds the planet's magnetic field with the potential to put it out of phase jeopardizing the integrity of the planet.[33]

---

[33] Zickuhr, Clare, Smith, Gar, *Earth Island Journal* quoting David Yarrow, electronics researcher. www.ufoarea.com

The polar slots in the magnetosphere have widened from a norm of six degrees to 25-46 degrees. This allows more matter and energy radiating from the Sun's solar wind and from interplanetary space to enter the polar regions causing the Earth's crust, the oceans and the ice caps to warm. Low frequency radio waves are vibrating the magnetosphere and causing high-energy particles to cascade into the atmosphere. Just by pushing a button and inducing these frequencies we can alter the signals in the Van Allen radiation belts, setting the stage for weather anomalies. Power line harmonics (resonance from high-tension wires) already cause the fallout of charged particles from the Van Allen belts to create ice crystals, which precipitate rain clouds. The planet is currently experiencing many weather anomalies such as 2004-05 record winter high temperatures in Greenland causing rapid melting of the ice pack. Is this just a result of man-induced global warming or are we adding to the recipe for destruction?

The depletion of the ozone layer by natural and man-made factors may have a new perpetrator. The high-frequency emissions from HAARP alter the temperature of the ionosphere. Although the operators state the ionosphere returns to normal upon cessation of the "ray," chemical reactions that produce ozone have already been altered. Our efforts to enact laws to protect the ozone layer may be useless if the government controlled ionospheric heaters prove to be many times more destructive to the ozone layer in a shorter period of time. This is a case where an "oops" may not be able to be fixed. Non-uniformity in the earth's ozone concentrations cause abrupt growth in temperature gradients creating more wind and wet weather. In the USA, the winter of 2005 saw many more days of wind with velocities in excess of 20 knots and record-breaking precipitation, especially in southern California.

Global warming is not the only threat to the planet's survival. The earth is a huge magnetic sphere. This magnetic field permeates and contributes to all life as the backbone of our atmosphere. The field varies in strength and consistency through the ages.

Significant shifts in the magnetic poles contributed to the theories of a major polar shift that wiped out the dinosaurs and earlier civilizations. A significant shift by the magnetic poles began in 1998. Normally we see a trend to a wandering magnetic pole field, as much as 3 km per year over ten years. Recent data shows the Arctic magnetic pole has traveled towards eastern Siberia by more than 120 km during the period between 1973 and 1984. Polar shift acceleration leads scientists to believe we are headed for an inversion of the magnetic poles with an increase in acceleration expected to 200 km per year. This shifting is requiring many airport runways to be renamed as their magnetic orientation is changed. There is a significant increase in world magnetic anomalies independent of the Earth's main magnetic field. It is the inversion of the magnetic fields process, which is causing the various transformations of the geophysical process. The acceleration of the pole movements are influenced by highly charged material from interstellar space, which have broken into the interplanetary area of our Solar System. This is not only affecting the magnetic poles, but the ozone content distribution and the increased frequency and magnitude of significant catastrophic climactic changes. Just as the Ancients experienced a similar set of interplanetary conditions 10,000 years ago, we are faced with a déjà vu situation.

Should a shift occur whereby the north and south poles moved to the equator, a major reorganization of continents and civilizations would occur. One theory is that Atlantis is the continent of Antarctica and the top security that reduces public exploration is to protect the secret. Although the theories claim a pole shift every 20-30,000 years, if the poles did shift 10,000 years ago, then it is very plausible that this once temperate continent now lies under the ice.

During the previous twenty-five million years, the frequency of magnetic inversions was twice in half a million years. For the last one million years we have seen eight to fourteen inversions. Scientist figure we are 26,000 years from the last pole shift with another predicted for the year 2012. A typical movement seems to be about a 1,000 mile shift in a few days. It is not the planet that shifts but the thin outer skin that lies over the molten core. It is suspected that a vibrational upset triggers a change in the rotation of the planet. Something big, like an asteroid, could generate this type of repercussion. Electrical motors and generators can be caused to wobble when their circuits are affected. Could human activities cause a significant change in the planet's electrical circuit or field and become the trigger? "While changes in the Earth's electric field resulting from a solar flare modulating conductivity, may have only a barely detectable effect on meteorology the situation may be different in regard to electric field changes caused by man-made ionization.[34] It may be that the Ancients in their quest for world dominance unleashed their harshest frequency weapon. This could have created a resonance that compounded itself, imploding with other frequencies to create rogue frequencies that pulsed at a rate equal to the waves created by a meteor hit. It could have been the "rapid pulse" that created Earth's heart attack.

    The trigger may also have come from inside the earth. Tesla's experimentation with tapping the earth's electromagnetic energy and sending signals through the core to far away destinations may not have been his smartest invention. If, in fact, he was a reincarnated Atlantean he may have not remembered that this technology could have precipitated the downfall of that civilization. Part of the technology scheduled to be implemented at the HAARP facility, is the ability to penetrate beneath the Earth's crust over most of the Northern Hemisphere. This would permit the detection and precise location of tunnels and other subterranean installations as well as improve communications with submarines. As we saw in the discussion on the Atlantean power systems and Tesla's wireless power systems, electromagnetic waves spread around the globe almost instantaneously from a lightning bolt or man-induced similar power source. Could these subsurface electromagnetic signals send a

---

[34] ibid

similar reverberation throughout the mantle of the planet disturbing the geologic equilibrium?

The earth basically is a huge magnetic sphere. Living things have been tied to its natural magnetic field for all time. Human use of electromagnetism for power and communications has produced an abnormal electromagnetic environment unlike anything we have experienced in 10,000 years. Epidemiological and laboratory studies have proven man-made frequencies are alien to the human being, resulting in profound health risks. Not only are they harmful to our very being, but to our planetary pulse as well. All electromagnetic fields are force fields, carrying energy and capable of producing an action at a distance. They have the characteristic like sunlight of having both waves and particles. Our man-made EM fields are alien to the planet's magnetic field and are creating a field of disharmonics that cannot be healthy for Mother Earth. The field frequencies created by the Ancients may have been responsible for agitating stable planetary pulses and along with CMEs and planetary alignments that competed for gravitational pull, the planet may have simple given up and tipped over.

Interplanetary influences on the earth have been logged and predicted by the Ancients. Supposedly, the Atlanteans were forewarned by their astronomers that a major earth shift was coming. We, too, are being warned in the texts laid down by our ancestors and the similarity in interstellar changes that are now affecting Earth. According to the Mayans, in 2012 we will experience a winter solstice that coincides with a rare astronomical alignment that only happens once in 26,000 years (similar to the time of the last pole shift.) Our Sun will gradually bring its position into alignment with the very center of our Galaxy (the galactic equator). Our solar system is moving towards the central bulge along the galactic equator just below the Dark Rift, which we suspect is filled with dust and planetary debris. This transition across the equator began in 1998 and should exit the area on the other side of the Dark Rift in 2018. Many predict a shift in man's consciousness during this transition.

From the decoded Egyptian glyphs, the story of Osiris (Orion) relates mathematical calculations from the year 10,000 BC that predicted a former cataclysm in 21,312 BC and had forecast the "coming" cataclysm to be July 27, 9792 BC, which is when the Atlantean civilization was destroyed. The Mayans also mathematically calculated that on December 21-23, 2012, the same astronomical conditions would exist to create the potential for another pole shift and cataclysm. Mythological texts constantly refer to "the sun fell into the sea" and "the sky is coming down." This would be indicative of a rotational reversal where the sun previously set over land in the east is now setting over the ocean in the west. Support of this theory comes from scientists who have found fossils of sea creatures high in today's mountainous regions and skeletons of sea mammals and animals indigenous to warm climates discovered in the inland arctic areas.

The question arises – what is behind these polar reversals? As shown above, interstellar changes are occurring, possible because of our solar system's current position within the Milky Way. The Atlanteans (although on the other side of the Milky Way equator during the cataclysm) saw a correlation between the magnetic field of the sun and the looming disaster. Both the Ancient and Mayan texts have been interpreted to indicate a reversal of the magnetic field of the sun causing super flares (CMEs) where trillions of particles would descend on the earth's poles. Normally rebuffed by the magnetosphere, this incessant bombardment will slowly etch away our defenses and streams of radioactively charged particles will leak through overcharging the Van Allen belt and finding their way into the electrojet. This continuous stream of electromagnetism would overcharge the Earth's fields generating unknown electrical forces that could trigger the change in rotation. Although the sun does go through magnetic reversals more often than every 10,000 years, the interstellar particles may not have been triggered by an enormous CME. We are currently seeing an increase in these streaming ions into the polar regions of our planet. It appears the scenario is being set with or without the sun flip-flopping.

If we do experience a pole reversal the effects will immediately create potential differences that instantly burn out all electronics and electric motors. All communications would cease. Aircraft would become giant gliders. Unfortunately, modern aircraft glide like rocks. The result to a frequency-based civilization would be disastrous even if geological events didn't wipe us out. This scenario lets us know how frighteningly dependent we are on technology and how vulnerable we are to cosmic forces.

The one missing question for which I have not discovered an answer – did the cosmic changes trigger the pole reversal or did man play a part? Obviously cosmic power surpasses our technological inventions tenfold or more. But, what if the normal transitions through space create these ion particle assaults that are just a part of our solar system's history? What if anomalies surfaced but no polar reversal occurred under normal circumstances? What if there was an external manipulative trigger that either weakened our planetary atmospheric defense system or exacerbated the electrical imbalance?

If we look at the power grid in place during the time of the Ancients, we can see potential for disrupting the Earth's natural harmonics. Using the ionosphere as a conductor and the earth as a power source could have widened the polar cusp angle allowing an influx of interplanetary particles. If we correlate that to present-day technological misuse, we, too, may be widening the "slot" with ionospheric heaters located near the polar regions and with subterranean radio-frequency assaults.

Chapter VI:
Wave Technology:
The big threat to modern civilization

Even if the cosmos is kind to us, we may face a bigger, more immediate threat to our civilization. As described in previous chapters many of the "star wars" weapons are now in fact reality. Most use what can be called wave technology or frequency based devices. One of these, EMP (electromagnetic pulse) has the ability to devastate the electronic/electrical infrastructure of a country. Can you imagine our highly technological society being leveled to a pre-electrical era? Automobiles, airplanes, boats, computers, telephones, power plants, light bulbs, air conditioning, heating, well pumps, elevators and anything with wiring will instantly cease to function. What a great terrorist victory that would make!

    A nuclear blast detonated 500 miles above a country would create an EMP that could paralyze civilization below it. This detonation generates massive outputs of x-rays and gamma rays, some of which will interact with the air molecules of the upper atmosphere. The result is a pulsed current of high-energy electrons that will interact with the earth's magnetic field. Instantaneously an invisible radio frequency wave is produced, one that is a million times stronger than the most powerful radio signal. This wave would reach everything in its line of sight at the speed of light. These high-speed electromagnetic "shock waves" would endanger much of our technological infrastructure and disable or permanently destroy all unprotected devices with circuits, wires, microchips, etc. Following this initial surge would be a less intense EMP that would assault electronics still left operating, followed by a third wave that would disrupt currents in electricity transmission lines, damaging the surviving electrical supply and distribution systems.

"Some foreign analysts, judging from open source statements and writings, appear to regard EMP attack as a non-lethal weapon, because EMP would inflict no or few immediate civilian casualties. EMP attack appears to be a unique exception to the general stigma attached to nuclear employment by most of the international community in public statements. Significantly, even some analysts in Japan and Germany – nations that historically have been most condemnatory of nuclear and other weapons of mass destruction in official and unofficial forums – appear to regard EMP attack as morally defensible. For example, a June, 2000, Japanese article in a scholarly journal, citing senior political and military officials, appears to regard EMP attack as a legitimate use of nuclear weapons."[35] Nuclear weapons are available to terrorists and to countries with a grudge to bear. The missiles needed to launch them are available from countries (such as North Korea) for $100,000. This makes the threat even greater.

It is also easier to disrupt a country's infrastructure on the local level. The original theory for a non-nuclear EMP producing device, thought up in 1927 by Dr. Arthur Compton to study atomic particles, makes use of injection of plasma into low electron count elements. By the mid 1980s, scientists had found ways to build a high-energy device that, without resorting to a nuclear blast, could emit a huge EMP. A one time explosive device provides kinetic energy required to rapidly build an electromagnetic field through electromagnetic induction rather than through the nuclear chemistry found in a nuclear explosion. A second, low-cost technology uses a moving short in a tube fed by a charging system. This technology, known as FCG – Flux Compression Generator – turns out to require far less cash to develop and manufacture. The mechanical construction of the FCG is actually quite simple – a college graduate in electronics or physics can accomplish an effective design of such a device. It consists of an explosives-packed tube placed inside a slightly larger copper coil. The instant before the

---

[35] Pry, Dr. Peter Vincent, EMP Commission Staff, Before the U.S. Senate Subcommittee on Terrorism, Technology and Homeland Security, March 8, 2005

chemical explosive is detonated, the coil is energized by a bank of capacitors, creating a magnetic field. The explosive charge detonates from the rear forward. As the tube flares outward it touches the edge of the coil, thereby creating a moving short circuit. The pulse that emerges makes a lightning bolt seem like a flashbulb by comparison.

This EMP bomb is only effective in a finite area about the device. The larger the armature of the device, the larger the electromagnetic field produced. Thus a device could be one foot across and take out very localized equipment, say a control facility or communications system. A device four or five feet across could be used to take out all communications at an airport or from a skyscraper take out the semiconductor devices for several miles in a swath extending out in all unshielded directions. Protecting equipment by surrounding it with a Faraday Cage (similar to screening in that which is to be protected) may not be an effective defense. Very-high-frequency pulses, in the microwave range, can worm their way around vents in Faraday Cages. If the equipment survives the first wave, it still may not survive as the outside electrical systems that feed the devices may have been affected causing electric surges to travel through the power and telecommunication infrastructure.[36] Terrorists would not have to drop an E-Bomb directly on a target – just on the key electrical generating sites or telecommunications centers.

The U.S. military has been embracing this weaponry with gusto. The Navy wants to use the E-bomb's high-power microwave pulses to neutralize antiship missiles. And, the Air Force plans to equip its bombers, strike fighters, cruise missiles and unmanned aerial vehicles with E-bomb capabilities. Countries such as India and several other Asian nations are working on both devices and countermeasures. Other nations savvy about EMPS are China, Cuba, Egypt, Iran, North Korea, Pakistan and Russia. Most countries regard EMP attacks as information warfare rather than conventional warfare that kills people directly.

---

[36] Rohit Khare Rohit@KnowNow.com,Thu, 3 Jan 2002 20:37:47 -0800

One company, Plasma Sciences Corporation, has developed a potential defense for electronics: a plasma limiter that provides protection from EMP bursts, and directed RF energy weapons. This system effectively protects electronics from short duration high power pulses. Unfortunately it is not in place worldwide so most countries are still highly vulnerable to this disabling "non-lethal" weapon. In 2005, the Senate Judiciary Subcommittee on Terrorism, Technology and Homeland Security concluded, "The 9-11 commission report stated that our biggest failure was one of 'imagination.'[37] No one imagined that terrorists would do what they did on Sept. 11. Today few Americans can conceive of the possibility that terrorists could bring our society to its knees by destroying everything we rely on that runs on electricity. But this time we've been warned, and we better be prepared to respond."[38]

Unfortunately, hardening systems is difficult and expensive. Not only must the device be shielded, but also antennas and power connections must be equipped with surge protectors, windows must be coated with wire mesh or conductive coating, and doors must be sealed with conductive gaskets. Fiber optic cable is not vulnerable to EMPs, but the switches and controls that use microelectronics in conjunction with the fiber optic cable will need to be protected. Each of these protective measures must be maintained on a regular basis. This is a massive task and as of today, we are virtually 'naked' to EMPs. We also need a firm policy in place for emergency procedures in case of attack. A massive population without communication, access to food, water, heat or cooling can turn into a very ugly crowd.

When we look at the weapons used by ancient technologically we find a striking similarity to those presently being created. Plasma-based weaponry is any group of weapons designed to use high-energy ionized gas or "plasma", typically created by superheating lasers or superfrequency devices. One plasma prototype weapon exists in Russia that focuses beams

---

[37] 9-11 Commission Report, 11.1 Imagination, p. 339
[38] Buzzle Staff & Agencies, "The Real Threat of Electromagnetic Pulse Weapons," www.buzzle.com 10/1/2005

of electromagnetic energy produced by laser or microwave radiation into the upper layers of the atmosphere. These beams would be able to defeat any target flying at supersonic or near-sonic speeds in the near future. A cloud of highly ionized air arises at the focus of the laser or microwave rays, at an altitude of up to 50 kilometers. Upon entering it, any object – a missile, an airplane, is deflected from its trajectory and disintegrates in response to the fantastic overloads arising due to the abrupt pressure difference between the surface and interior of the flying body. What is fundamental in this case is that the energy aimed by the terrestrial components of the plasma weapon – lasers and antennas – is concentrated not at the target itself but a little ahead of it. Rather than "incinerating" the missile or airplane, it "bumps" it out of trajectory.[39] When I wrote the novel, *2012 Airborne Prophesy*, I incorporated this technology into the story in order to expose the possibly misuse of the weapon. One can ionize the atmosphere by just heating up it up with electro-magnetics (laser, microwaves, etc.), but it is a whole lot easier to generate plasma if one dumps barium, carbon, and aluminum into the atmosphere first.[40]

PEP (pulsed energy projectiles) is another high-tech weapon created to produce pain and temporary paralysis. The result of an electromagnetic pulse produced by expanding plasma that trigger impulses in nerve cells, resulting in stimulating the pain neurons without damaging tissue. This weapon had been scheduled for operation in 2007. Another unthinkable weapon is genetic poison that will probably be used in the future as the most optimal variant to destroy a part of the population, even a certain specific part. Each human race has an individual genetic code. Genetic differences may lay the ground of the genetic weapon. It will be possible to create genetic viruses to exterminate a certain group of people on the planet. Bioelectronic weapons may be

---
[39] new scientist: http://www.newscientist.com/article.ns?id=dn7077
[40] "Horizontal plasma antenna using plasma drift currents," 6,118,407
  Navy patent

coming to a country near you. Professor G.Bogdanov patented a radiation generator to fight termites. The generator's radiation kills insects paralyzing their nerve system. Can we be assured that tweaking this device won't affect humans as well?

Weather wars may already be in progress. It is known that Russian scientists have been working in the field of the meteorological war in the city of Obninsk. In 1967 the U.S. Senate passed the Magnusson Bill authorizing the Secretary of Commerce to accelerate programs of applied research, development and experimentation in weather and climate modification. At that time experiments were or were scheduled to occur in 22 countries. These included the airborne rain seeding programs, fog clearing, hailstorm abatement, forest fire control and lake storm snow redistribution. By 1973 there were over 700 degreed scientists and engineers in the U.S. whose major occupation was environmental modification. In October, 2005, *Newsweek Magazine* reported that China had 35,000 people engaged in weather management and modification. Weather wars were imminent. Therefore, in 1978 the United Nations Convention on the Prohibition of Military or Any Other Hostile Use of Environmental Modification Techniques (ENMOD) prohibited the use of techniques that would have widespread long-lasting or severe effect on a region caused by weather manipulation.[41] Numerous countries signed this agreement, including the United States, but research and testing by non-governmental groups continued as a private (permitted) venture. Some groups suspect that the erratic pattern of hurricane Katrina may have been the result of an electromagnetic weather war from several of these independent groups.

As we mentioned before in the discussion regarding HAARP, it is possible to alter specified weather on a particular territory by changing the electric charge of the air. This facility sends radio frequencies 72,000 more powerful than the largest single AM radio station into a specific spot at the very top of

---
[41] Smith, Jerry E., "Weather Goes To War," *Atlantis Rising*, No. 65, Sept/Oct 2007

the atmosphere, the ionosphere. They heat this spot by several thousand degrees controlling and directing the processes for what they call ionospheric enhancement. Can they use this technology for weather manipulation? The technology suggests it is possible, but that is not the stated aim of HAARP. Hard rains, droughts, blizzards can cause very serious damage to an enemy or a country that is a target for takeover. Could some of the severe weather being experienced by the world today a result of intentional weaponry or just a by-product of us mucking around with the upper atmosphere?

Can man-triggered earthquakes trigger tsunami's halfway around the world? There is consideration given to the theory that underground nuclear testing or electromagnetic pulses targeted below the earth's crust can have a definite effect on fault lines in other parts of the world. This, too, could be construed as a potential weapon. Nothing is easier to camouflage than generating a natural disaster that immobilizes a target population.[42]

Laser weapons are available today. They initially dissipate slowly, but this changes the further away the target is. Laser weapons cause 100% damage at point blank range, 100% damage at short range, 75% at medium range, 25% at long range, and 10% damage at extreme range.

Particle beam weapons dissipate slowly overall and are more effective at most range categories because of the nature of their energy beam. Particle weapons cause 100% damage at point blank range, 100% damage at short range, 100% at medium range, 75% at long range, and 50% damage at extreme range.

Plasma weapons dissipate quickly as the super-heated plasma quickly cools outside of the weapon. Plasma weapons cause 100% damage at point blank range, 50% damage at short range, 25% at medium range, 10% at long range, and 5% damage at extreme range.

Other energy weapons, including ion weapons, dissipate at a relatively constant rate. These energy weapons cause 100% damage at point blank range, 100% damage at short range, 75%

---

[42] http://fathersergio.wordpress.com

at medium range, 50% at long range, and 25% damage at extreme range.

Obviously, we know that these weapons are in use today, otherwise the statistics would only be hypothetical. Each of these weapons carries an energy signature. Once fired that energy wave imparts an effect on surrounding air molecules generating a cascading effect. Energy doesn't go away. It just changes form. Where does it go? Do we know if this affects the energy pulse of Mother Earth? These are questions that may not be answered, nor do the creators care to test the full impact on the total energy wave interaction with the pulse of the planet. Responsible technology would examine all the scenarios and hazards before implementing a device. Unfortunately, similar to the drug industry, recalls may be in order once the negative implications are proven. In the case of unseating the pulse that keeps our planet stable, that may be too late.

According to Investigator James Russell, "Recent observation suggest a connection with global change in the lower atmosphere and could represent an early warning that our Earth environment is being changed."[43] NASA's Aeronomy of Ice in the Mesosphere (AIM) spacecraft has revealed ice crystals forming in the summer at altitudes of 50 miles over the polar regions. These clouds have been forming more often and at lower latitudes in recent years, leading some researchers to speculate that there is a link in the high-altitude changes they represent and changes at the lower altitudes attributed to global warming. Could these changes from the upper altitudes be exacerbated by electromagnetic pulses being directed from Ionospheric Heaters? And if so, are they accelerating the change in our Earth's environment?

The question raised concerns about the ability of technological advances to explore all the side effects of these new devices before they become a hazard to our environment.

---

[43] Russell, James, 3rd, Hampton University, "NASA's Aeronomy of Ice in the Mesosphere," *Business & Commercial Aviation*, Aug. 2007

If we look at ancient history, we conjecture that the science of the time never forecast the destructive capability of their power systems and weapons as they applied to the overall stability of the planet. Once unleashed without restraint it was too late to fix and Earth's polarity was compromised. We do know that there is an electromagnetic pulse to the operation of the planet. And we are aware that they pulse is changing. The magnetic pole is shifting and the atmosphere is becoming more unstable as evidenced by the increase of severe weather. Are we tickling Mother Earth and is she beginning to react? Are we creating a reoccurrence of the disaster that befell our ancestors and with the same triggering mechanism – electromagnetic pulses and frequencies saturating the airspace? Is there hope? If we look at the prophecies and if we recognize that the human thought pattern also radiates frequencies, then just maybe we can change the outcome.

Chapter VII:
Can we change the prophecies?

We take a look at the prophecies outlined in the Bible Codes, Hindus, Hopi Indians, seers and the Mayans to understand the similarity of their warnings. Many of these prophecies have been handed down by translators and through myths. This leaves the original objective of the prophet open to interpretation. The story of Noah's Ark or a similar great flood tale is told by religious factions all over the world as well as remote tribes that have little contact with our material world. The fact that the legends seem to have no continental boundaries indicate that something actually happened in history that affected the entire planet – a great flood. The warning that we will experience another apocalypse may be nothing more than a retelling of past stories. According to cosmic law, the planet goes through periodic upheavals. I cannot discount that, and another pole shift may, in fact, be around the corner.

But by looking at the various prophecies we may see a similarity in the prophets' objective. Most deliver a spiritual message – that a disregard for Mother Earth and separation from the natural order of things will bring death and destruction to civilization.

Could these prophets have been warning us not to do what the Atlanteans did? Or, could they be trying to control our behavior by evoking fear into our hearts giving us incentive to comply with their "laws?" Just suppose that a technologically superior alien race inhabited Earth and wanted humans as their work force, nothing more. How would they keep them subservient – by making them believe that technological advances would doom them to an eventual apocalypse? As civilizations drifted from spirituality towards materialism, many forgot that they were earmarking their future to a disastrous end. Technology became their God and materialism their vocation. The Aliens had lost control over their subjects. Is this what is happening today? Maybe we weren't invaded by Aliens, but merely had descendants of a superior race (the Giants which were thought to precede the Ancients) trying to run the planet. Were the prophecies written to invoke fear and compliance generation after generation? Let's examine some of the prophecies to find the underlying theme.

Interpretations of the Bible code predicted major earthquakes in 2006 for Japan and Los Angeles, California. Asteroids that are part of a comet complex from the Sun's dark companion star threaten earth every 1,500 years. The last one registered an impact in 534 AD. The Bible code warns that asteroids from fragments of this comet swath will hit the sea near France in 2006 and in Russia and India in 2010, possibly triggering a pole shift. So far these predictions have been false. An asteroid designated 2002-NT7 two kilometers in diameter has actually been forecast by astronomers to impact the earth on February 1, 2019. The Bible Code also predicted a supervolcanic eruption in the western USA in the area of Yellowstone National Park. A massive eruption occurred in this area 640,000 years ago, spewing ash over the Midwest to the eastern Pacific. Beneath this area magma stretches fifty kilometers and is giving signs of a forthcoming eruption. The region north of Yellowstone Lake has bulged upwards by almost a meter in the last fifty years and hundreds of small earthquakes are registered each year. Should Yellowstone blow, it would produce a volcanic event 1,000 times greater than any ordinary volcano. The ash fallout could affect global climate and growing cycles.

Hindu prophecies explain that the age of Kaliyuga, in which we are now living, is the age of darkness and ignorance. It is the most degenerate and fallen of the ages. Characterized by a hardening of the spiritual core of mankind, an almost total lack of sacredness and a dedication to extreme materialism, this age will be closed by Kalki the 10th avatar of Vishnu

(Buddha was the 9th) as he ushers in a golden age of peace, prosperity and harmony.

In the book *Gospel of the Essenes*,[43] a translation of old manuscripts that existed in both Aramaic and Old Slavonic, we find reference to the end days. "– and the sixth angel sounded. And the heaven departed as a scroll when rolled together. And over the whole Earth there was not one tree, Nor one flower, nor one blade of grass. And I stood on the earth. And my feet sank into the soil, soft and thick with blood. Stretching as far as the eye could see and all over the Earth was silence." The angel said that men have created their own destruction, but says there is always hope. "What had always been, what was now and what would come to pass. I saw the holocaust that would engulf the earth. And the great destruction that would drown all her people in oceans of blood. And I saw too the eternity of man and the endless forgiveness of the Almighty. The souls of men were as blank pages in the book always ready for a new song to be inscribed. And in the middle of the river stood the Tree of Life. Which bore fourteen kinds of fruits, and yielded her fruit to those who would eat of it, And the leaves of the tree were for the healing of nations."

The Mayan calendar ends on December 21, 2012. The Maya lived in what are now southeastern Mexico and northeastern Central America where they constructed large cities and had an advanced knowledge of math, writing and astronomy. Their advanced society may have procured this knowledge because they were direct descendents of the Atlanteans. Their writings accurately depicted the Earth's rotation around the sun in 365.2422 days, just two seconds longer than time measured by today's atomic clocks. They also wrote about planets that were thought to be undiscovered in that age, planets like Neptune and Uranus. Adding to the mystery was their sudden disappearance 800 years ago. The Maya recognized ages of the sun (approximately 5,128 years)

---

[43] Szekely, Edmond Bordeaux, *The Gospel of the Essenes*, 1974, C.W. Daniel Co., Ltd.

that occurred with a Great Age of 26,000 years. These ages had a cycle of birth and death and periodically cleansed the planet of its civilizations, which had denounced spirituality in favor of materialism that threw life forces out of balance. The final phase of the Mayan calendar began on July 26, 1992 and forecast a moment in human destiny when time would speed up dramatically. Many of earth's present population are currently experiencing the feeling of not enough time to get things done. The Maya forecast 2012 to be the end of the Fifth Age and more importantly the end of the overall Great Age.

Interpretation of the end of time referred to in the calendar has led some Mayan followers to believe that this is the end of Planet Earth. Many also believe that the Mayans were giving us hope of entering into a more golden age where our bodies would be less dense and lifespans extended. Unfortunately just like the Bible, there were several versions of the Mayan calendar, the original and the "edited" transcription by Diego de Landa, a Franciscan from Spain who was successful in getting his works published and distributed widely. The de Landa version did not include information from the Mayan record keepers and much of the text was conjectured from de Landa's own conclusions. It is this version that we see quoted most often. The original Cholqij is more accurate and stresses the importance of their depiction of time, different from our current Gregorian calendar and how our future depends on our living within the rules of planetary and cosmic cycles.[44]

We also make reference to a period in history known as Harmonic Convergence as defined by Jose Arguelles, "the point at which the counter-spin of history finally comes to a momentary halt, and the still imperceptible spin of post-history commences." It was the fulfillment of the prophecy of Quetzalcoatl who lived at the very end of the first millennium. The prophecy of Quetzalcoatl states that following the ninth hell;

---

[44] Arguelles, Jose, "Harmonic Convergence From Prophecy to the Fourth Dimension. Navigating by the Law of Time," Indigo Sun Magazine online, 1998, www.indigosun.com

humanity would know and experience an unprecedented New Age of Peace. His prophecy of the Thirteen Heavens and Nine Hells is based on the Mayan time knowledge. After the destruction of the Mayan time knowledge the Vatican instituted a calendar "reform" in 1582. It is this Gregorian calendar (12 months/ 60 minute hours) by which we live our lives today. Why was it changed? Were the "rulers" of Earth trying to "reengineer" human actions by alienating them from the natural harmonics of the cosmos replacing it with a time frame that refocused life around a daily work schedule? The Gregorian calendar minimizes our ability to see the natural cycles and flow with them. It is designed for material functionality not independent thinking and feeling. It binds civilizations into a confined, structured environment where information can easily be censured and dissidents eliminated.

As Arguelles explains, the Harmonic Convergence was an announcement of the forthcoming end of time as we know it (Gregorian). Time is the mathematics of the universal laws of nature, the unifying force that holds everything together. The Mayans knew this using a frequency of time as the 13:20 ratio (13 moon, 28 day calendar [364 day/year with one extra day]) on which the Mayan calendar is based – considered natural time. The Harmonic Convergence began August 17, 1987, and is slated to end in 2012. Since the initiation of the Harmonic Convergence, there have been measurable increases in the energy of our planet from a resonance of eight cycles per second existing for thousands of years to twenty-three cycles per second today. Accompanying this energy boost is a rapid decrease in the magnetic field that surrounds Earth. At the end of the Harmonic Convergence we will again return to the original reference of time and into fourth dimensional awareness.

Various seers and channeled readings by psychics lead us to similar conclusions focusing on the opportunity for us to change consciousness from a dense third dimension to a more etheric fourth dimension. Many psychics offer the theory that Earth exists in several planes at once but in different time continuum.

WORSE THAN GLOBAL WARMING

Based on the reality of the previously mentioned Philadelphia Experiment's time shift, we cannot discount this possibility. In the book, *"Conversations with God, Book 3,"*[45] author Neal Donald Walsch offered God's words, "I am saying that once before on your planet you had reached the heights – beyond the heights really, to which you are now slowly climbing. You had a civilization on Earth more advanced that the one now existing and it destroyed itself. Not only did it destroy itself, but it nearly destroyed everything else as well. It did this because it did not know how to deal with the very technologies it had developed. Its technological evolution was so far ahead of its spiritual evolution that it wound up making technology God."

During the last century, an old wise woman of the Cree Indian nation, named "Eyes of Fire," had a vision of the future. She prophesied that one day, because of the white mans' greed, there would come a time when the earth being ravaged and polluted, the forests being destroyed, the birds would fall from the air, the waters would be blackened, the fish being poisoned in the streams, and the trees would no longer be, mankind as we would know it would all but cease to exist. There would come a time when the "keepers of the legend stories, culture rituals, and myths and all the Ancient Tribal Customs" would be needed to restore us to health making the earth green again. They would be mankind's keys to survival. They were the "Warriors of the Rainbow." There would come a day of awakening when all the peoples of all the tribes would form a New World of justice, peace, freedom and recognition of the Great Spirit.

The ancient Hopi Indian prophecy states, "When the Blue Star Kachina makes its appearance in the heavens, the Fifth World will emerge." The Hopi name for the star Sirius is Blue Star Kachina. They gave us signs of the future and the prophesy.

The first sign: We were told of the coming of the white-skinned men. Men who took the land that was not theirs and who struck their enemies with thunder. (Guns)

---

[45] Walsch, Neal Donald, *Conversations with God Book 3*, 1998, Hampton Roads Publishing, VA

The second sign: Our lands will see the coming of spinning wheels filled with voices. (Covered wagons)
The third sign: A strange beast like a buffalo but with great long horns will overrun the land in large numbers. (Longhorn cattle)
The fourth sign: The land will be crossed by snakes of iron. (Railroads)
The fifth sign: The land shall be criss-crossed by a giant spider's web. (Power and telephone lines)
The sixth sign: The land shall be criss-crossed with rivers of stone that make pictures in the sun. (Concrete roads and their mirage producing effects)
The seventh sign: You will hear of the sea turning black and many living things dying because of it. (Oil spills)
The eighth sign: You will see many youth who wear their hair long like our people, come and join the tribal nations to learn our ways and wisdom. (Hippies)
The ninth and last sign: You will hear of a dwelling place in the heavens above the earth that shall fall with a great crash. It will appear as a blue star. Very soon after this the ceremonies of the Hopi people will cease. (Asteroid)

The Hopi Prophecy story also said that the signs of the third-world shaking would be given when: trees will die (acid rain and demise of the rainforests); man will build a house in the sky (space station); cold places will become hot and vice-versa (global warming and erratic weather); lands will sink into the ocean and lands will rise from the sea; the appearance of the Blue Star Kachina. It is reported that all the signs have been fulfilled and we are destined for an apolcalypse.
 As we examine the prophecies, we see the underlying theme that something bad would happen if man didn't obey. The bad thing may have differed depending on who was

telling the story, but they all stressed that Mother Earth and our spiritual beliefs were not to be destroyed. Of course, what have we done in the last millennium? The environment is suffering. As a race, humans do not respect other life forms and maintain a strong belief in their own superiority. Spirituality is slowly making a comeback and there are growing factions that are fighting for a return to the feminine, softer approach to life. But, technology is the dragon and he is extending his fiery breath into all corners of our civilizations. He is making it impossible to exist without the convenience of electricity, cell phones, computers, drugs, television, homeland security, weather manipulation and teleconferencing. Tell a cell phone user that it has not been proven the device is safe for your health and see how easily they discount the information. Ask children to go play outside and see how quickly they make excuses to stay indoors and surf the net. We are rapidly being alienated from the harmonic frequencies of the planet and instead becoming slaves to a man-made, frequency-based society.

Technology is not bad. From my research I've determined that it's just being misused. There is power in technological superiority. That power could be misused if we ignore looking at the bigger picture of its effect on our planet. Just as we have discovered that our disregard for the environment has advanced global warming, will we find that our love affair with electronics, microwaves, wireless technology and star wars weapons create an unbalanced electromagnetic field that will unbalance our planet's rotation and cause a global apocalypse? As we saw in the ancient texts, civilizations focused only on the power they achieved with technological advances. They, like us, never considered the bigger environmental picture. Are we following in their footsteps by ignoring the writing on the ancient wall? Will we, too, precipitate a polar shift?

The Atlanteans amassed an arsenal of weapons that could have destroyed galaxies. How many nuclear bombs and earthquake machines do you need to conquer a few countries with similar weapons? Arsenals in the U.S. stockpile include nuclear weapons, electromagnetic frequency generators that not only burn victims but can alter thinking processes, lasers, ionospheric high-frequency arrays for weather manipulation, stealth aircraft, earthquake generators, methods of distributing lethal viruses en masse, and the list goes on. We have the ability and the "firepower" to destroy everyone on the planet. Just like the Atlanteans, the USA has environmental policies that ignore implications of future repercussions. The Europeans are doing a better job and trying to rectify their assaults on Mother Nature, a Rama quality. Other nations like China will eventually compete with the USA for global superiority and at that time we may see the reenactment of the Atlantean/Rama war of 10,000 years ago.

    If we take heed in the prophetic warnings, by ignoring the laws of nature we will certainly meet our apocalyptic destiny. Yet, we have an opportunity to change the outcome. Political action can be effective if those in power will listen. But according to the prophecies, Hindu teachings, and new frequency-based research, we may be able to create a new hologram, a new vision. The Philadelphia Experiment did a lot to open up our minds to the possibility of alternate realities or other dimensions. It is scientifically agreed that atoms, electrons, protons and quarks are mostly made up of space – nothingness that is frequency-based. They arrange themselves according to their velocities. This is true of emotion as well where thoughts create a wave-length that can affect the health of our bodies as well as our feelings. Matter is made up of molecules, which are made up of atoms. Atoms are comprised of leptons, fermions, etc which are made up of photons which contain quarks. Matter has a low frequency. Mind has a high frequency. The ultimate constituents of all matter are luminiferous ether (sometimes called space having many properties). These fine particles are sensitive to thought (consciousness) or mind being sympathetic to the frequencies of each.[46]

    The magnitudes of time and space are nothing but modes of

---

[46] Pond, Dale, "Sympathetic Vibratory Physics – It's a Musical Universe," 12/9/2000, www.svpvril.com/Cosmology

thought depending upon a point of view. They exist only relatively to your perception. Although it is difficult for gravity-bound humans to grasp, the dense materialistic world we have created is made up of air and frequency. Blood has its own rhythm, as does gasoline. Both are liquid and both are made up of vibrations. Just as a hologram can be reproduced in total by just taking a small part of the image, we can manifest much in our life by creating thought vibrations in our own mind. We exist in a world of effects arising from an unseen cause. Each invention began as a thought vibration. We can create fear by repeated media portrayal of terrorism that can blossom into creation of a "protective" police state complete with border guards, biometric scans at all public places and data bases full of our personal information. We have taken a thought vibration, spread it through the media where individuals embraced it and created more synergistic vibrations until the whole country pulsed with fear and manifested the need for protection. Basically we created a hologram (vision in all our minds) that became reality.

Self-help gurus teach us how to envision success; health and wisdom claiming the hologram will manifest if we stick to our vision. "Thoughts are things" is a belief system that has been around since sayings were passed down. Churches use prayer which can heal the dying and books have been written about thoughts manifesting and healing disease. If vibrations can be so powerful to create our lives then surely they can change the future. A hologram is nothing more than an image of something that is perceived as real but is actually nothing. The prophecies create a world that has gone away from spirituality and man must meet the final punishment. If the masses believe this we will see the end days materialize. On the other hand if the vibrations are of a higher level and the preservation of Earth's pulse is maintained, the hologram will reflect harmony and a shift in consciousness and time rather than a cataclysmic apocalypse. In the year 2000 we saw a media frenzy about Y2K. Corporations spent billions preparing for this non-event. Large numbers of people bought into this and stocked up on food, rewrote wills and prayed. The rest sat and watched going about their daily lives as if nothing would change – and nothing did. The hologram manifested the wishes of the many who thought nothing would change, but for the few, maybe their computers crashed.

Further research must be done quickly on the effect our lower vibrational thoughts are having on the pulse of Mother Earth. The electromagnetic fields that are saturating our atmosphere are having an effect on humans and the planet. Remember when airplanes ceased to fly for the few days post 9/11? How peaceful it was. Vibrations changed, calmed down and besides the quiet we noticed a subliminal quieting of our body's pulse. Since then air travel has resumed to pre-9/11 levels and frequency saturation of the magnetosphere has returned to unprecedented levels. The pulse of our human body and of the planet cannot rest. We are both defending ourselves and trying to achieve balance. If political change is not possible we can change the hologram. We may even be able to change the dimensional future of our planet to ward off cosmic assaults.

2012 does not have to see a cataclysmic end of the world. It can mean a change in the time line to one where we live harmoniously and a higher vibration with the Earth and one where greed and power have been eliminated from our consciousness. There is a global movement in this direction that is gaining momentum. We do not have to repeat ancient history and destroy ourselves to gain wisdom. We just have to open our eyes and learn from our mistakes.

Epilogue

*A moon drenched sea
 caresses my ship with gentle undulations.*

*Alone… a microscopic spec
 in a vast wilderness embracing quiet thought.*

*I sense a universal wholeness,
 a purpose for being knowing no confinement.*

*We are all together surviving on a planet
 precariously flourishing in an alien environment.*

*What meaning has one life
 as an isolated drama?*

*Dangers to the self, emotional or physical,
 hold importance, but for an instant
 as one man's crisis becomes another's news story.*

Time becomes history without permission;

We are building blocks sharing the same space.

There are no enemies.

*Each life exists as an education,
 powerful if shared,
 not individually, but as a sense of evolution.*

Nina Anderson, 1984
composed at an altitude of 35,000 feet

Bibliography

9-11 Commission Report, 11.1, Imagination; P. 339

A. Gorbovsky, *Riddles of Ancient History*, 1066, Soviet Publishers, Moscow

Arguelles, Jose, *Harmonic Convergence From Prophecy to the Fourth Dimension. Navigating by the Law of Time*, Indigo Sun Magazine online, 1998, www.indigosun.com

Begich, Nick and Roderick, James, *Earth Rising II*, Earthpulse Press

Berlitz, Charles, *The Bermuda Triangle*, 1974, Doubleday, NY, 2003

Budge, Sire E.A. Wallis (translator), *The Queen of Sheba and Her Only Son Menyelek* (Kebra Nagast), 1932, Dover, London

Buzzle Staff & Agencies, "The Real Threat of Electromagnetic Pulse Weapons," www.buzzle.com, 10/1/2005

Childress, David Hatcher, & Clendenon, Bill, *Atlantis and the Power Systems of the Gods*, Adventures Unlimited, Kempton, IL, 2000

de Caro, Chuck, *The Zap Gap*, The Atlantic, March 1987, p 24-28

Geryl, Patrick & Ratinckx, Gino, *The Orion Prophesy*, Adventures Unlimited Press, IL, 2001

Gordon, Wade, *The Brookhaven Connection*, 2002, Sky Books, New York

Jacoby, Jeff, "The jury is still out on global warming," *The Boston Globe*, August 20, 2007

Jochmans, Joseph Robert, *Time Capsule: The Search for the Lost Hall of Records in Ancient Egypt*, Alma Tara Publishing, SC, 29731

Just Looking, photography by Lewis Whyld for South West News Service, *Smithsonian* magazine, May 2004, p28-29.

Keith, Jim, *Mass Control: Engineering Human Consciousness*, IllumiNet Press, GA 1999

Lockhart, Gary, *The Weather Companion*, Wiley and Sons

Nichols, Preston & Moon, Peter, Pyramids of Montauk, Sky Books, NY, 1995

Noorbergen, Renee, *Secrets of the Lost Races*, 1977, Barnes & Noble Publishers, NY

Pickert, Kate, "The World Turns To Desert," *Popular Science*, p.48, August, 2007

Pond, Dale, *Sympathetic Vibratory Physics – It's a Musical Universe*, 12/9/2000, www.svpvril.com/Cosmology

Rifkin, Jeremy, *The European Dream*, E Magazine, March/April 2005, p34

Rifkin, Jeremy, *The European Dream: How Europe's Vision of the Future is Quietly Eclipsing the American Dream*, Tarcher/Penguin, 2004

Roy, Chandra Protap (translator), *The Mahabharata*, Calcutta, India, 1889

Sanderson, Ivan T., Investigating the Unexplained, Prentice Hall, Englewood Cliffs, NJ, 1972

Szekely, Edmond Bordeaux, *The Gospel of the Essenes*, 1974, C.W. Daniel Co., Ltd.

Smith, Jerry E., HAARP, *The Ultimate Weapon of the Conspiracy*, Adventures Unlimited, 1998

Tesla, Nikola, *Colorado Springs Notes*, 1899-1900, Aleksandar Marincic, Nikola Tesla Museum, 1978

Thomas, Andrew, *We Are Not the First*, 1971, Souvenir Press, London, UK

U.S. Explores Russian Mind-Control Technology, Defense News, Jan. 11-17, 1993

Wall, Judly, "Timeline: Electromagnetic Weapons," *Resonance Newsletter*, posted on: Raven1.net, Mon Sep 3, 2007 (referemce p. 51 of text)

Walsch, Neal Donald, *Conversations with God Book 3,* 1998, Hampton Roads Publishing, VA

Zickuhr, Clare, Smith, Gar, *Earth Island Journal* quoting David Yarrow, electronics researcher, www.ufoarea.com

2012 Airborne Prophesy

blends high technology, ancient catastrophe, political chicanery and new age concepts to give us an adventurous story sure to have readers seriously examining their options for the future. Join in as our heroine's corporate jet takes her into a mysterious atmospheric rift that could rain environmental havoc down on all corners of our Earth as we experience the impact of terrorism and the very real technologies government and big business are employing to combat it.

hard cover $ 23.95
soft cover $ 16.95

"Corporate Pilot Nina Anderson knows a good mystery when she sees one, and with her background in natural healing and a talent for telling stories, she has crafted a winner with *2012 Airborne Prophesy*."
Women in Aviation magazine, July, 2004

"I was amazed at some things the author mentioned in the book that I had never stopped and actually thought about before. It was wonderful! All in all, this is a great book that blends modern fiction, sci-fi, and just a bit of romance."
Huntress Reviews, excerpted review from Detra Fitch

"What a wonderful book. At first I didn't know what to expect, but the further I read, the more I got interested. I couldn't put it down. It was like reading something we know is happening today."
M. Burrett, Newspaper Reviewer, Nashville, IL

More titles from
Safe Goods/ATN Publishing
www.safegoodspub.com *or* **888-628-8731**

Cancer Disarmed Expanded	$ 7.95
Spirit and Creator, *The Mysterious Man Behind Lindbergh's Flight to Paris*	39.95
Spirit Print (suitable for framing)	39.95
The Backseat Flyer	9.95
Eliminating Pilot Error	7.95
Human Factors and Pilot Error Video	19.95
The ADD and ADHD Diet	10.95
Cell Towers, *Wireless Convenience or Environmental Hazard*	19.95
The Natural Prostate Cure	6.95
Lower Cholesterol Without Drugs	6.95
No More Horse Estrogen	7.95
Testosterone Is Your Friend	8.95
Overcoming Senior Moments	9.95
ADD, *The Natural Approach*	4.95
The Smart Brain Train	7.95
Kids First: *Health With No Interference*	16.95
Analyzing Sports Drinks	4.95
2012 Airborne Prophesy	16.95
Protector	19.95

Author: Nina Anderson

Nina Anderson is a corporate jet pilot, a Certified Specialist in Performance Nutrition through the International Sports Science Association, and author of 17 books, including *2012 Airborne Prophesy, Cancer Disarmed, ADD, The Natural Approach* and *The Backseat Flyer*. As a nationally acclaimed author, alternative health expert, television and radio personality, Nina's message has reached thousands of listeners. She has been an active researcher in the alternative health field for over twenty years and actively examines metaphysical, paranormal and scientific anomalies.

Nina holds a BA from Monmouth College, currently flies Hawker corporate jet aircraft and teaches safety seminars for the FAA Wings Program. Her experience flying at high altitude triggered her keen interest in atmospheric issues and frequency-based systems of communications and weaponry. She has spent well over ten years amassing the information presented in this latest work, ***Worse Than Global Warming.***